MANAGEMENT OF TOXIC MATERIALS IN AN INTERNATIONAL SETTING

A CASE STUDY OF CADMIUM IN THE NORTH SEA

COASTAL WATER SERIES:

1. *Water Quality Management Plan North Sea: Framework for Analysis,* Rob Koudstaal, 1987.
2. *Management of Toxic Materials in an International Setting, A Case Study of Cadmium in the North Sea,* Lies Dekker, Blair T.Bower & Rob Koudstaal, 1987.

DELFT HYDRAULICS
P.O.Box 177, 2600 MH Delft, Netherlands

Management of Toxic Materials in an International Setting

A Case Study of Cadmium in the North Sea

By

LIES DEKKER, BLAIR T.BOWER & ROB KOUDSTAAL

A.A.BALKEMA / ROTTERDAM / BROOKFIELD / 1987

Cover design: Bureau Paul Mijksenaar (Sandra Neerincx)

Published by
A.A.Balkema, P.O.Box 1675, 3000 BR Rotterdam, Netherlands
A.A.Balkema Publishers, Old Post Road, Brookfield, VT 05036, USA

ISBN 90 6191 795 6

Printed in the Netherlands

Preface

This report is one of the products of the Analyzing Biospheric Change (ABC) Program sponsored by the International Federation of Institutes for Advanced Study (IFIAS). The ABC Program was established in 1981 with the objective of adding to knowledge of how to manage natural resources such that exploitation of those resources can be done while simultaneously maintaining or improving ambient environmental quality and living standards. The ABC Program is oriented to decision making at the regional level and to empirical analysis. The Program consists of three "problem areas": utilization of coastal waters; urbanization; and agriculture in semi-arid areas. This report is a product of the first problem area.

Management of coastal waters in many areas involves international settings. In developing a set of case studies for the coastal waters problem area, a necessity was to include case studies involving such settings.

An opportunity consistent with this objective was presented when the Dutch government, in May of 1982, presented an official note to the Dutch Parliament entitled, "Harmonization of North Sea Policies" (Ministerie van Verkeer en Waterstaat, 1982). The note initiated a study to formulate a Water Quality Management Plan for the Dutch portion of the North Sea. In the note the following terms were used with respect to the plan:

> Building on international conventions and actual national policies, the plan will represent the vision of the Dutch government on a coherent and strategic approach to preventing and combatting pollution of the North Sea. The plan will provide directives for international negotiations and in addition will indicate management strategies as far as they can be taken in the Netherlands.

The involvement of Delft Hydraulics in both the ABC coastal waters problem area and in the Dutch government study enabled testing the framework for analysis developed in the problem area in a real analysis situation. The existence of the technical study by the Dutch government provided a background for exploration of the institutional problems of managing the North Sea.

Technical details such as discharges of materials and their effects were taken from the Dutch North Sea study. Many of the figures used herein are reproduced from that study. Permission to use this material is much appreciated. Information on and outputs from the Dutch study were provided by Paul Baan, Peter de Bruyn, Peter Glas, Hans van Pagee, and Henk Visee; each is associated with Delft Hydraulics. Dr.L.D. Guruswamy, of the Faculty of Law, University of Durham, United Kingdom, participated in the early discussions of the institutional problems. Typing, drawing and final editing was taken care of by Nel Woldhuis and Engelbert Vennix.

Table of contents

1 Introduction

1.1 BACKGROUND AND OBJECTIVES

Two primary considerations spawned the ABC North Sea study. One was the international context of marine resources management of the North Sea, a situation characteristic of many coastal waters and marine resources management regions in the world. The international context of the North Sea provided a setting for exploring how nations individually, and through an international mechanism --- in this case the European Community (hereinafter referred to as EC) --- cope with a resource utilization problem relating to a common property resource.

The other was the initiation by the Dutch government in 1983 of a study to develop a Water Quality Management Plan for the Dutch portion of the North Sea (hereinafter referred to as WQMP-North Sea). The Dutch study to development a plan was carried out in 1983 and 1984. The study concentrated on the waste disposal use of the North Sea and was mainly set up to support existing governmental policies, both on the national and international level, and to formulate --- if required --- additional directives for future governmental actions with respect to toxic materials entering and accumulating in the North Sea. The concern for waste disposal and water quality management on the part of the Dutch government was reflected particularly by a concern for heavy metals.

The framework for analysis developed in the ABC problem area, succinctly described in the Appendix A, has a much wider scope than the Dutch study. In the first place the ABC framework includes all human activities using marine resources. A second, more important, difference is involved in the concept of a plan. The Dutch plan results in so-called policy directives. No specific management actions were considered. The concept of a plan in the ABC approach includes specific physical measures installed at specific times in specific locations and considers specific incentives to induce users of such physical measures to adopt desired behavior in order to produce a specified time pattern of outputs. In such a concept of a plan, and of planning, different targets should be considered, for example for production levels or concentrations of polluting substances in the North Sea, and the information to be provided on aspects such as costs, benefits, and administrative consequences, should enable the selection of targets and corresponding measures by decision making entities.

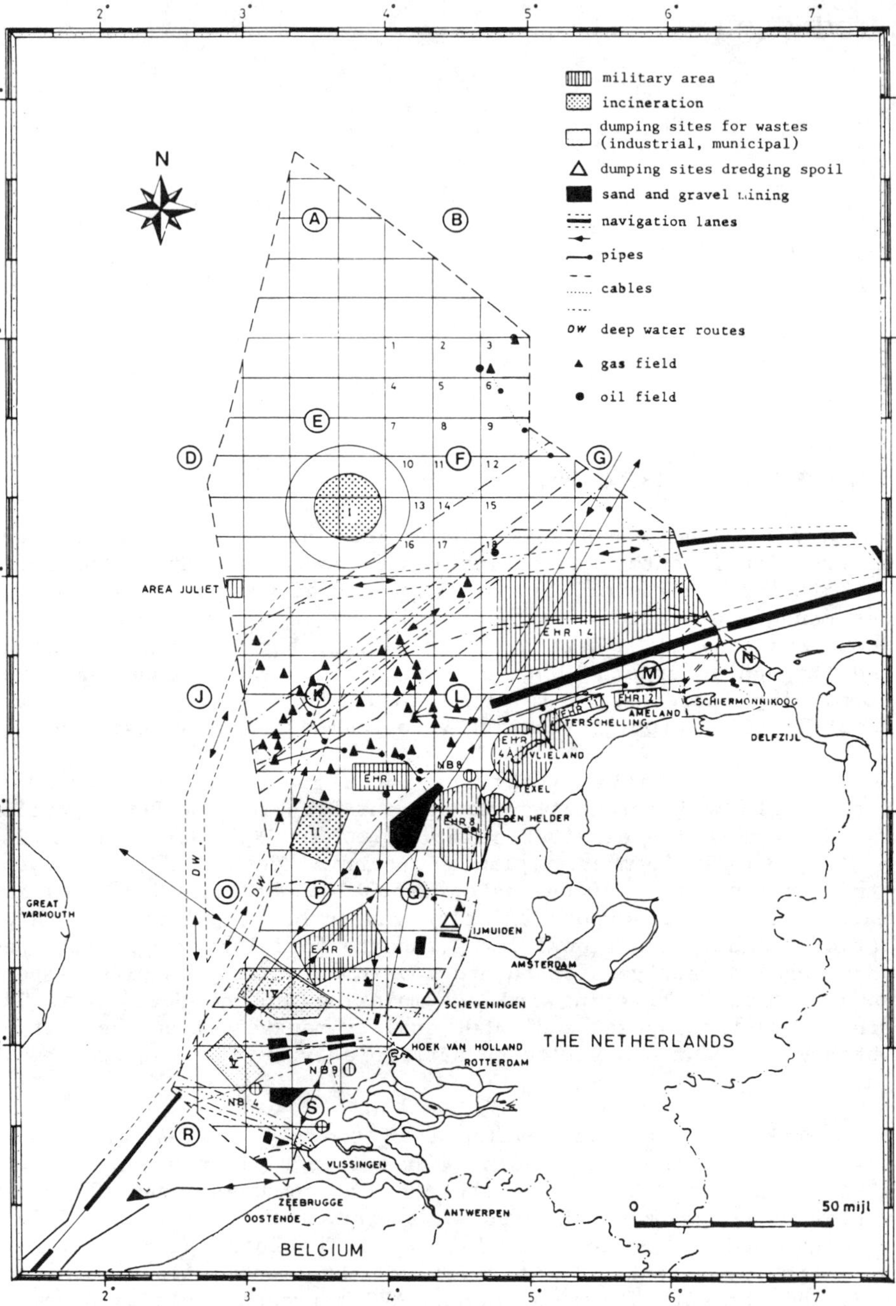

Figure 1-1. Overview of some activities on the Dutch portion of the North Sea

The ABC study presented herein has to be considered as a modest "ex post" evaluation concentrating on only one of the toxic substances dealt with in the Dutch study, namely cadmium. Consistent with the overall ABC objectives, the ABC North Sea study is oriented towards the following two kinds of audiences and related general objectives. To analysts involved in planning for management of marine resources, lessons will be drawn on the confrontation of the ABC coastal waters framework for analysis with a study carried out in a real world situation. To decision making agencies, the study wants to show what kind of information an integrated analysis can generate in addition to the results of the Dutch WQMP-North Sea study. In this dualistic approach, the ABC study should be considered as one further step aiming at providing better and more relevant information which, through a proper understanding by decision making entities of the possibilities and limitations of such information, finally would lead to better planning and management of marine resources.

More detailed information on the focus of the ABC study and the motivation for the selection of cadmium is given in Chapter 2, which includes as well the description of the organization of this report. In order to make the report self containing, some information is included on the natural system of the North Sea and on the context of North Sea management. This is done in the remainder of this chapter, which concentrates on general North Sea management aspects and gives a short description of the main features of the natural system.

1.2 MANAGEMENT OF THE NORTH SEA

The North Sea is a body of water and associated shoreline, a marine resources region, which can provide --- and is used to provide --- multiple outputs of products and services. These outputs include: marine transport; sand and gravel; exploration for, and production of, oil and gas; commercial fishing for shellfish and finfish; recreational boating and fishing; habitat for resident and migratory birds and marine animals; marine sanctuaries, i.e., where no other uses are permitted; waste disposal, both direct and indirect; and areas for defense activities. Some of the uses of the North Sea are shown in Figure 1-1.

Some outputs produced in a marine resources region have adverse impacts on, and in some cases preclude, the production of other outputs. Thus, marine resources management involves making decisions with respect to resource utilization questions such as: allocation of special zones for exclusive use by one or more activities, e.g., marine sanctuaries, navigation lanes, military zones, offshore mining zones; location of dumping sites for industrial wastes, sludge, dredging spoil and the amounts and timing of dumping; amounts of sand and gravel to be extracted where at what times of the year: and amounts of finfish and shellfish to be caught where at what times of the year. This decision milieu defines a basic problem in the management of a marine resources region, namely, determination --- and then the implementation of the production --- of the "optimal" mix of outputs of products and services over time. For a detailed description of marine resources management, see Bower, et al. (1982).

The determination of the optimal mix must be made and implemented in the context of several basic facts.

One, although the North Sea has long been partitioned among the bordering countries with respect to areas for national oil and gas exploitation, for other uses the North Sea represents a common property resource "owned" by all the bordering nations. Because the traditional approach to exploiting common property resources, i.e., "open access", has often lead to excessive exploitation and consequent domination or elimination of certain outputs, some institutional mechanism is required for management to prevent such a result. Waste disposal is one use which can easily result in excessive exploitation.

Two, some of the resources and uses of a marine resources region are mobile, such as migratory fish species, marine transport, and dispersion of wastes discharged into the water body. Thus, cross boundary flows into and out of the region occur and must be considered explicitly in management.

Three, closely related to two is the fact that a significant portion of the demands for products and services of the North Sea stem from beyond the boundaries of the countries bordering the North Sea. Marine transport is a prime example. Liquid wastes entering the North Sea from the south via the English Channel represent another example. Therefore, analysis for management decisions must explicitly consider factors affecting demands external to the region.

Four, demands for products and services vary over time, within a given year, from year to year, and over longer periods of time. For example, demands for recreational uses occur primarily during the summer season; demands for disposal of liquid wastes occur all year, but at varying levels. The demand for drilling for, and production of, oil and gas offshore varies from year to year in relation to economic conditions. The mix of outputs desired from a marine resources management region will change over time as a result of:

- changing interests of society, as interpreted by governmental authorities, and as reflected in legislation, policy statements, rules and regulations, guidelines, plans, and programs;
- increasing knowledge of the basic physical/chemical/biological processes in marine ecosystems and of the interrelationships between and among the activities producing the various products and services in a region;
- changing technology of producing products and services;
- changing prices of factor inputs to production;
- changing values of products and services in relation to other sources of the same or substitute products and services (changing demands);
- increasing knowledge of the quantity and quality of resources available; and
- changing export and import policies, implementation of international conventions.

Five, there are uncertainties in terms of the short-run and long-run effects of disposal of wastes on ambient water quality, including quality of sediments, and in turn uncertainties with respect to the effects of the changes in water quality on various components of the marine ecosystem, e.g., subaqueous vegetation, fish. The effects of waste disposal, e.g., on a particular species of fish, are a function not only of the quantity, quality, time pattern, and spatial pattern of disposal of wastes, but are related to and interact with: meteorological and hydraulic conditions; reduction in habitat as a result of dredging, dumping, and land reclamation; and fishing pressure as reflected in effort and catch over time.

1.3 NATURAL SYSTEM OF THE NORTH SEA

The North Sea is a rather shallow shelf sea with an average depth of less than 100 meters. The boundaries and tributary areas of the North Sea are indicated in Figure 1-2. A distinction can be made between a southern part and a northern part, the boundary being the 56 degree north parallel. The southern part of the North Sea has an average depth of only 37 meters. Depths of more than 200 meters are only found in the northern part in front of the Norwegian coast.

Winds and tides introduce a more or less regular water circulation. Water movement in the southern part is mainly directed towards the northeast. In the central part an anti-clockwise circular movement occurs, as shown in Figure 1-3. Streamlines indicated in this figure show residual, i.e., average over the tidal period, water flows of 25,000 cubic meters per second (m^3/s). Areas with thermal or salinity stratification are limited and are found mainly in the deeper parts in front of the Norwegian coast. Retention times are long. Water entering the southern North Sea through the English Channel leaves the northern boundary in front of the Norwegian coast after 150-250 days. The shorter retention times are in winter, because of stronger, prevailing southwesterly winds.

Substances entering the North Sea are transported according to the stream pattern shown in Figure 1-3, either in solution or adsorbed on suspended material. The suspended material accumulates mainly in estuaries in the southwestern part of the Netherlands and along the English coast, and in an extensive intertidal sedimentation area beginning north of the Netherlands with the famous Wadden Sea area and continuing eastward through the German Bight to the Danish west coast. Another sedimentation area is found in the deep northern part of the North Sea.

1.4 USE OF NORTH SEA FOR DISPOSAL OF WASTES

The North Sea provides a sink for disposal of wastes from sea-based activities, such as marine transport, and for wastes produced by a diversity of land-based activities. The wastes from the latter enter the North Sea via river inflows, direct discharges or dumping of liquids and solids, and atmospheric deposition. Disposal of wastes --- both directly and indirectly --- is an important use of the North Sea. This has long been the case for all countries bordering the Sea. Direct disposal involves discharge of a waste directly into a water body, e.g., sewage treatment plant outfall and material contained in dredging spoil dumped in the sea. Indirect disposal refers to

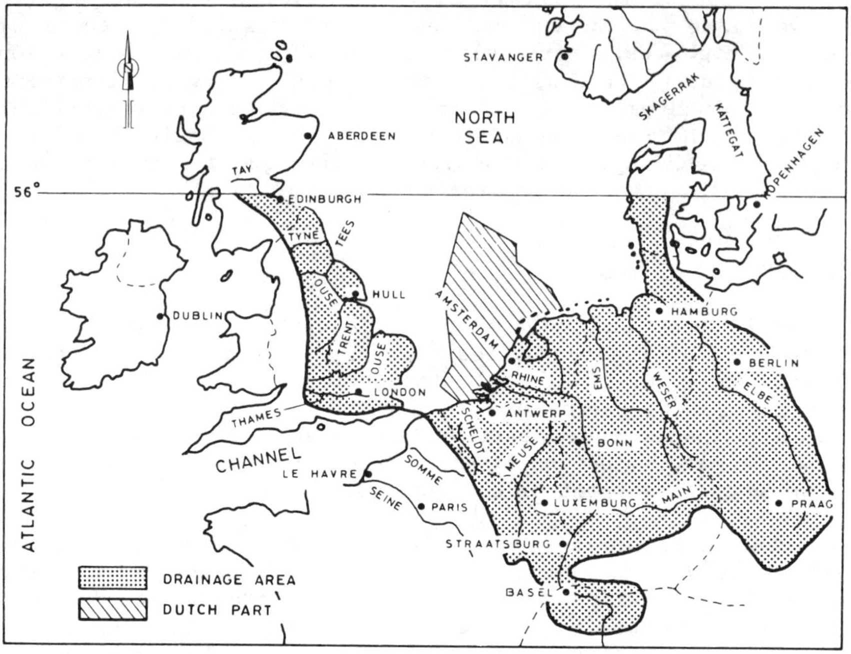

Figure 1-2. Boundaries and tributary areas of the North Sea

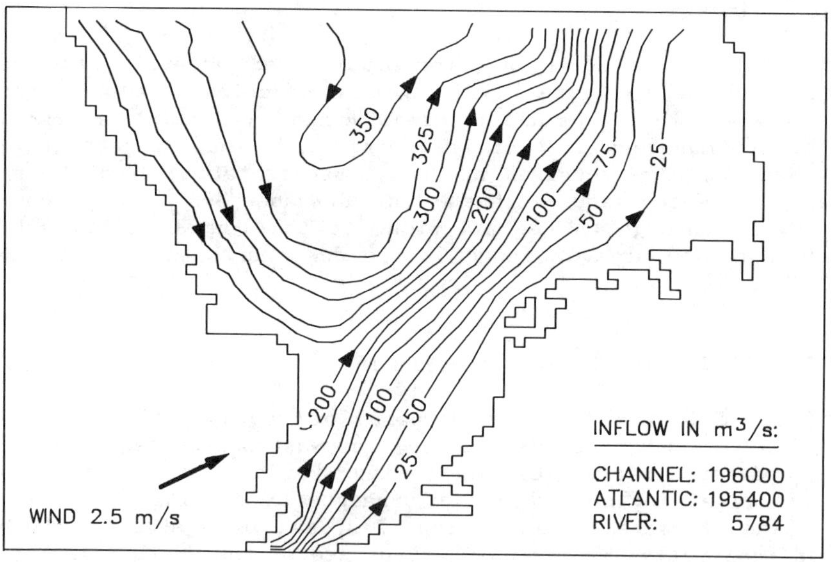

Figure 1-3. Water movement in the North Sea

materials which are discharged into a medium other than water, such as the atmosphere, which subsequently reach the North Sea via dry or wet deposition from the atmosphere. Combustion of coal in power plants and incineration carried out out on ships offshore are examples of indirect disposal.

As populations and economies in the countries tributary to the North Sea continue to grow, more wastes will be generated and pressures will increase for additional disposal into the North Sea, particularly as sociopolitical resistance to, and costs of, land disposal increase.

Concern for effects of waste disposal

The North Sea constitutes an highly productive ecosystem, providing habitat for both resident and migratory species. Various substances discharged, such as heavy metals, are taken up by algae and benthic organisms and are transferred to and accumulated in higher levels of the food chain. The following effects have been noted at the top of the aquatic food chain, and have been attributed to --- but not yet proven to be caused by --- the presence of toxic substances in water and sediments of the North Sea.

- Skin diseases have been found on flat fish in the German Bight and in the Dutch Wadden Sea.
- Locally, some massive fish kills have occurred.
- Between 1940 and 1965, very large reductions occurred in the populations of two bird species, the sandwich stern (Sterna Sandvicensis) and the elder (Somateria Mollissima).
- Significant reductions in the resident seal and dolphin populations occurred in the period since World War II.

In addition to these effects, substantial loads of phosphorus and nitrogen have increased the rate of algal growth in some areas, resulting in oxygen depletion in bottom layers and subsequently in changes in the abundance and diversity of species throughout the food chain in those areas. Changes in magnitude and composition of species in an aquatic ecosystem occur not only as a result of waste discharges, but also by loss of habitat and direct disturbance by human activities. The former, for example, occurs as a result of land reclamation and disposal of material, such as dredging spoil, on the sea bottom. Direct disturbance is exemplified by fishing activities.

Even if there were to be no increase in the discharge of substances to the North Sea, there would likely be a continuing buildup in the concentrations of certain substances in the sediments of the North Sea, as a result of continuation of the present, or even reduced, discharges. Depending on the extent to which the adsorption process is reversible, the quality in the water column will be affected to a greater or lesser degree by accumulation in the sediments. This fact was one of the factors leading to the initiation of the Dutch study of water quality in the North Sea.

Water quality management context

In a publication on the effective management of resources, Fotheringham and Birnie (1979) described the management context of North Sea marine pollution as follows.

Pollution in the North Sea is multinational in its origins and both transnational and national in its consequences. States consequently have demands to address to each other for the reduction of frontier-crossing pollutants which particularly affect them. In addition, international agreement on pollution control is seen as essential to a fair apportionment among commercial competitors of the costs involved.

Economic conditions are the primary determinants of political action, or lack of it, to "combat pollution". The economic problems of pollution reduction derive from the fact that the market has not operated to include the costs of protecting or cleaning up the "common property resources" of the North Sea. Including the costs of disposing of the "inevitable residual mass" generated in any activity may mean increasing the costs of production and therefore increasing selling prices. But governments typically are reluctant to impose constraints which might --- but do not necessarily --- increase costs and hence might reduce the competitiveness of their national economic units.

Particular pollution questions get on national agendas when public concern is sufficiently aroused as a result of pollution incidents. Normally, governments are subject to competing pressures from those potentially at risk from pollution and those whose businesses or jobs might be adversely affected by required preventive measures. The politics of pollution are dominated in most cases by the allocation of the costs of pollution reduction, rather than by the distribution of the profits of resource exploitation, such as with offshore oil production and harvesting of fish.

Because of the multiplicity of sources of discharges of pollutants, many governmental departments and agencies within nations and from various international organizations are involved in pollution politics.

1.5 LEGAL AND INSTITUTIONAL CONTEXT FOR MANAGEMENT OF THE NORTH SEA

Marine resources management in the North Sea must distinguish between two legal and institutional contexts, the territorial zones and the full North Sea. In territorial waters the coastal states execute full sovereign power, with the exception that free passage is to be given to "innocent" ships. In the full or open sea, the utilization of marine resources is --- or can be --- regulated through international conventions. Many conventions exist within different contexts, e.g., global, regional, European Community, and are related to different uses of the marine resources. Countries are free to join or not to join such conventions.

The conventions often lead to more regular cooperation among countries. Frequent meetings provide operational means for communication at a non- political level to implement the corresponding conventions. Activities within such cooperation include, for example: developing regulations; monitoring implementation of international agreements on national levels; and monitoring the effectiveness of the corresponding regulations.

In addition to the conventions and related or associated agreements, the European Community --- which does not include Norway --- plays an important role in the management of the North Sea, in particular regarding fisheries management and the integration of national environmental actions. The main instrument of the EC in the latter connection consists of directives, which generally result from previous coordination between and among the countries involved. In the case of Norway, cooperation is developed via negotiation between the EC and Norway.

All of the above mentioned conventions and EC directives have no operational authority in the sense of control and sanctions. However, the EC directives do have legal authority in the sense that EC countries are "obliged" to implement these directives in their national legislation and procedures. Little knowledge exists of how the directives are actually implemented in the individual countries, given that no formal sanctions exist for failure to carry out the obligation. This was one of the major questions to be addressed in the ABC study, as is discussed in the next chapter.

In addition to the fragmentation because of the international nature of the North Sea, responsibilities for marine resources management within each nation are divided among several or many governmental entities. None of the North Sea countries gives responsibility to a single agency for integrated management of marine resources; different ministries representing different interests are involved. For example, in the Netherlands the principal agencies involved at the national level are the ministries of: Transport and Public Works; Housing, Physical Planning, and Environment; Economic Affairs; and Agriculture and Fisheries. In 1977 an Interdepartmental Commission for coordination of North Sea Affairs (ICONA) was established, with the secretariat provided by the Ministry of Transport and Public Works. ICONA has no operational authority, but is considered an important institution to harmonize actions of the different ministries.

In addition to problems of coordination at the national level, coordination is critical, but often difficult, among national, provincial, and local levels of government. There is no formal coordinating mechanism, ala ICONA, to achieve coordination among these levels in the Netherlands.

2 Focus of the ABC study

In Chapter 1 the two stimuli for the ABC North Sea study, the international context of marine resources management of the North Sea and the Dutch WQMP-North Sea, have been introduced. Analyses supporting the WQMP-North Sea focussed on waste disposal use of the North Sea. Therefore, it was decided that the ABC North Sea study would also concentrate on waste disposal. With the WQMP-North Sea study as a base, ABC funds were used to focus on two issues stimulated by the Dutch study and fitting into the objectives of the ABC coastal waters problem area: (1) the kind of analysis to support planning which can be useful for decision making in an international setting; and (2) the triggers, effects, and implementation of regulatory measures in such a setting.

The ABC study had to decide on what waste disposal problem to focus. For example, the EC has listed 129 substances of concern with respect to their polluting effects, such as heavy metals and synthetic organic compounds. After explaining below why cadmium was chosen as the focus of the ABC study, the specific questions to be answered in the study are set forth. In the last section of Chapter 2, the organization of the study and of the report are indicated.

2.1 WHY CADMIUM

Cadmium was chosen for analysis because of: (1) increasing concern for potentially adverse, long-run effects of heavy metals such as cadmium; (2) the existence of an EC directive on cadmium, one of the six substances --- the others being mercury, lindane, DDT, pentachlorophenol, and carbontetrachloride --- for which directives on pollution of the aquatic environment have been adopted by the EC; and (3) the recommendation in the Dutch WQMP North Sea to pay additional effort to the elimination of cadmium discharges, on top of actions which already had been planned.

The heavy metal cadmium represents a class of substances with the following characteristics:

- it is toxic to many organisms exposed to relatively small concentrations, i.e., parts per billion;
- it is persistent in the environment after being discharged, i.e., it is not biodegradable; and
- it accumulates in organisms.

Adsorbed to suspended material cadmium accumulates in coastal and riverine sediments. Physical processes may cause a resuspension of sediments. Chemical processes may result in dissolution of adsorbed cadmium. Biological processes may also play a part in the "fate" after discharge.

Cadmium occurs in water in both dissolved and adsorbed forms. In fresh water 70% to 90% of cadmium is in the adsorbed form. In salt water the adsorbed fraction ranges from 0% to about 40% (Salomons, 1983). The major part of adsorbed cadmium in sea water is irreversibly bound. The distinction between adsorbed and dissolved cadmium is made because cadmium irreversibly bound to suspended particles is not biologically available, so that it is relatively non-toxic. Only the dissolved fraction of cadmium and its compounds can be considered toxic to aquatic organisms. The adsorbed fraction of cadmium and the speciation of dissolved cadmium, which influences the toxicity, depend on such factors as salinity, pH, and temperature. For example, when dredging spoils including sludge with cadmium bound to it are dumped into the North Sea, part of the cadmium dissolves as a result of the changed aquatic environmental conditions, e.g., higher dissolved oxygen content.

Aquatic organisms are exposed to cadmium in the water phase, in suspended material, in sediments, and in food. Cadmium can enter an organism by passing through the gills, by oral intake, or after contact with outer membranes. Cadmium also leaves the organism by the same processes. Metals such as cadmium are primarily taken up by aquatic organisms from the water phase, resulting in a form of bioconcentration called bioaccumulation. The resulting concentration in organisms depends on the chemical equilibrium of the toxic compound between the water phase and the organism. With respect to cadmium, bioaccumulation factors, i.e., the quotient of concentration in organisms and concentrations in water, have been found of 125 for grass shrimps and 10,000 to 40,000 for mussels. The toxic effect depends on the concentration inside the organism.

Information on the toxicity of cadmium for organisms is based on laboratory research. It has been found that cadmium in water influences energy management in organisms, resulting in less productivity, i.e., growth and reproduction, and an increased sensitivity to stress from natural factors. The growth of mussels decreases at cadmium concentrations in water of 10 milligrams per cubic meter (mg/m^3). Sublethal effects were found at cadmium concentrations as low as 1 mg/m^3. Lethal effects were observed in one experiment in which an hatch of Amphipod Pontorporeia Affinis died at a cadmium concentration in water of 6.3 mg/m^3 (Sundelin, 1983).

The results of laboratory experiments cannot be directly translated into behavior of organisms in coastal or marine waters. Field conditions cannot be replicated in the laboratory. In marine waters, a number of factors affect the health of the organism. In the real world environment, organisms are exposed to a mix of compounds, some of which may reduce, some may exacerbate, the effects of cadmium. For example, calcium appears to protect marine animals against the toxic effects of cadmium to some degree. Both the toxicity and the accumulation of cadmium are significantly affected by the ambient salinity, i.e., a decrease in salinity leads to an increase in toxicity, or the net uptake, of cadmium. Higher temperatures result in an increase in toxicity and/or accumulation of cadmium in marine biota.

Cadmium can also have toxic effects on humans. About one-third of the cadmium intake in humans goes to the kidneys. Sufficient research has been done to substantiate the fact that metals, and some synthetic organic compounds, have an affinity for particular parts or a part of the human body. In terms of quantity and effects, the kidneys represent the critical organ in humans with respect to cadmium. A cadmium concentration in the range of 200 to 300 mg cadmium per kilogram (kg) of kidney weight has been determined to be the critical level for the average human.

The mean intake of cadmium by which the critical concentration in the kidneys can be reached in 50 years is estimated to be on the order of 0.2 to 0.4 mg per person per day. Daily intake is a function of food consumption, smoking habits, exposure in the work place, ambient air quality. For example, smoking 20 cigarettes a day substantially increases the cadmium intake by the body (Stichting Natuur en Milieu, 1984). The World Health Organization has established a Provisional Tolerable Weekly Intake of 0.4 to 0.5 mg per person. In the Netherlands the mean intake of cadmium has been estimated to be 0.020 to 0.025 mg per day, or 0.140 to 0.175 mg per week, of which from 0.001 to about 0.120 mg per week are estimated to be via consumption of finfish and shellfish (Ministerie van Volkshuisvesting, Ruimtelijke Ordening en Milieubeheer, 1984).

2.2 THE ABC STUDY: QUESTIONS TO BE ANSWERED

As mentioned previously, the ABC study attempted to answer two kinds of questions. The first set of questions related to analysis in planning for water quality management, in particular to the kinds of information to be generated. The ABC study was particularly interested in costs of achieving different ambient water quality levels, with different geographic coverage, and the distribution of those costs. Such information is basic to understanding tradeoffs between (1) costs and (2) what is gained by use of such resources. That in turn leads to the investigation of the <u>worth</u> of the improved ambient water quality. The ABC study was not able to develop complete coverage on costs, and did not have time and resources to investigate the "worth" issue. However, the explication of the approach is useful in demonstrating the difference between the Dutch study and what is considered a study which would be more adequately related to decision making. In addition, the Dutch study gave no attention to implementation in relation to individual dischargers, a critical element in management, as discussed in Chapter 7.

The second set of questions explored the relations between, and interactions on, international and national levels. For example, does the stimulus to assess, and then to decide whether or not to do something about, a perceived ambient water quality problem originate from above, i.e., the international level, or from below, i.e., from one or more of the nations involved? What kind(s) of actions are involved at international and national levels and how effective are they?

These two sets of questions are reflected in Figure 2-1. This figure shows the steps contemplated in the evaluation of the framework for analysis and the empirical investigation of the effects of: the EC directive on cadmium; national laws of the Netherlands with respect to cadmium; and the interrelationships between the two levels.

Questions related to analysis

- What is the scope of the Dutch WQMP-North Sea, what kind of decisions had to be taken and would be the subject(s) of the plan, and how specific have these aspects been delineated?
- What steps of analysis were included, how were they executed, and how do they compare with the ABC framework? Is there a need for revising the framework?
- Was the information generated useful for the formulation of the plan?
- Is there any relation between the first question about the scope of the plan and the other questions about the steps of the analysis and the usefulness of the information generated?
- How have uncertain developments and lack of knowledge of the behavior of the North Sea ecosystem(s) been dealt with in the analysis? For example, given that atmospheric deposition was found to be a significant source of cadmium discharged into the North Sea, was any attempt made to determine how cadmium got into the atmosphere in the first place, i.e., before being deposited in the North Sea by dry deposition and wet deposition?

Questions related to institutional structure

Triggers

- What "triggered" the development and adoption of the international regulation, i.e., the EC cadmium directive? Was the international action begun prior to national actions, or did national actions/concerns stimulate the process at the international level? Is it likely that actions by a few --- 1, 2, 3 --- nations involved in an international body of water stimulate actions at the international level?
- To what extent did the concern for cadmium discharges into internal waters stimulate action in terms of the EC directive? Do countries, or some countries, use international decisions as a means to stimulate action internally in a direction desired by some groups within a country?

Effects of international regulations

- To what extent has the existence of an international regulation, i.e., the EC directive, affected the presence of cadmium in the international body of water, North Sea, and affected discharges into the North Sea from various sources?
- How do international regulations, e.g., the EC cadmium directive, get "translated" into national regulations and procedures?
- How are international regulations, e.g., the EC cadmium directive, implemented and enforced?

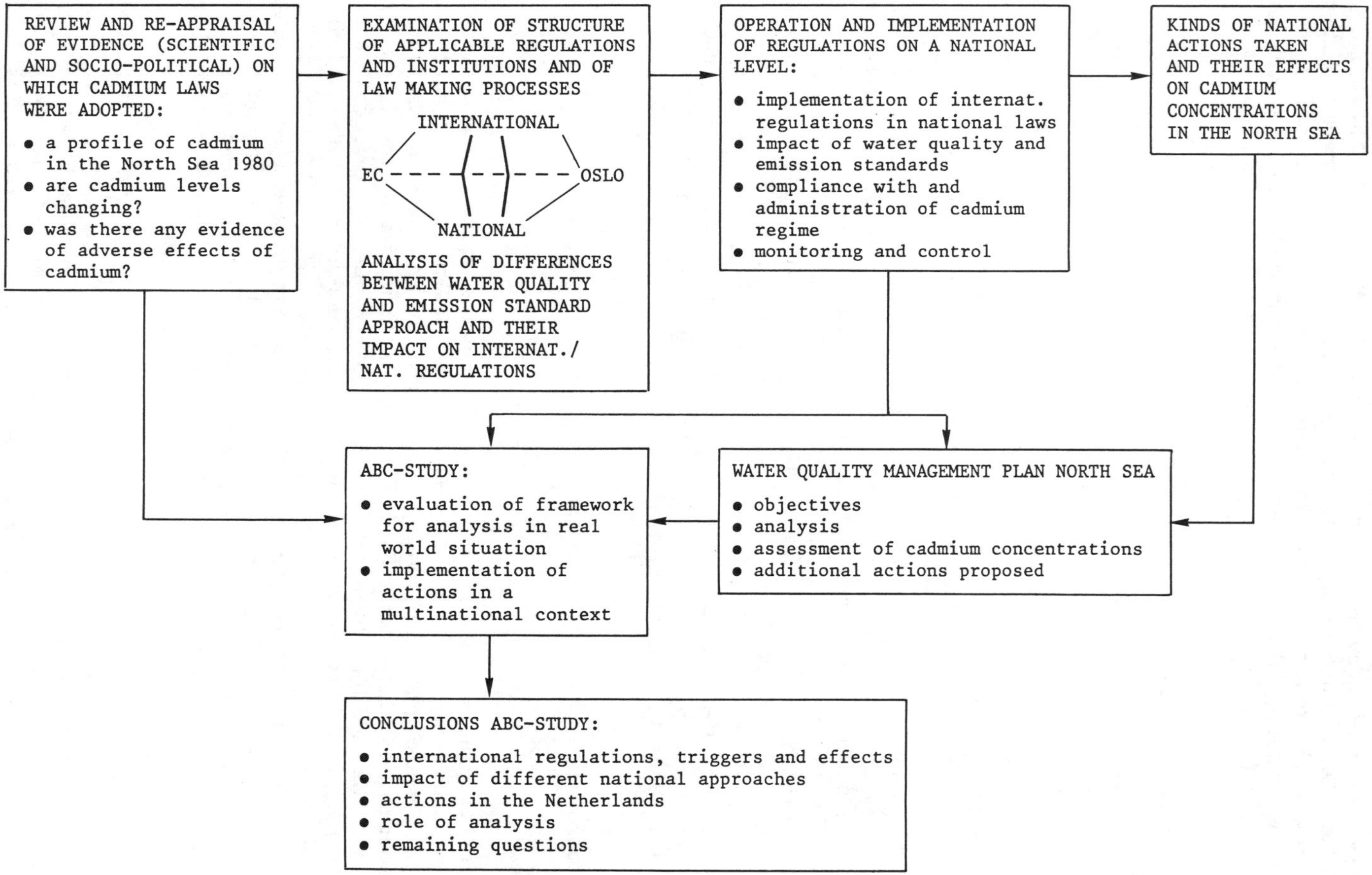

Figure 2-1. Sequence of steps in evaluation of framework for analysis and investigation of implementation of cadmium directive

- Is the EC cadmium directive sufficiently specific so that it can be implemented through national legislation/regulations/actions? Given that regional and local governmental authorities usually actually write and enforce permits, how is the EC directive "passed through" to the regional and local entities?
- How are discharge permits for individual activities, e.g., industrial plant, municipal sewage treatment plant, written in the various North Sea countries in relation to the EC cadmium directive? Is there any relation?

Implementation

- What are the scope and nature of incentives provided at national levels to dischargers, e.g., subsidies, permits, fines, technical advice, self-monitoring requirements, and random inspections?
- Has there been any attempt to define what behavior would be considered to be in noncompliance with the EC cadmium directive? Are there any sanctions which might be used to induce a noncomplying country, or dischargers within a country, to come into compliance? Or, do the levels set in the EC cadmium directive simply represent levels which all countries considered they could meet or were already meeting at the time the directive was negotiated?
- How does the uncertainty in terms of the long-run effects of cadmium --- or any other heavy metal or synthetic organic compound --- on the marine ecosystem affect the actions of the "parties at interest"? Are there different approaches to decision making under this type of uncertainty among the various North Sea countries? Are some countries more conservative, i.e., use larger factors of safety, than other countries? Are some countries more "risk adverting" than other countries? Are ambient water quality targets, and emission standards, adopted without analysis of their cost implications?

2.3 ORGANIZATION OF THE REPORT

To provide a basis for discussion of analysis procedure, a description of the Dutch WQMP-North Sea study is presented in Chapters 3, 4, and 5.

Chapter 3 contains the background of and objectives for the study, and the approach.

In Chapter 4, the methods for estimating discharges of cadmium in 1980 from the various sources are described. The estimated quantities are also shown.

The methods for estimating the effects of the 1980 discharges on ambient water quality and on quality of sediments, e.g., in terms of cadmium concentrations by season, are described in Chapter 5. Results under different assumptions are shown.

Institutional actions in relation to the set of questions posed in Chapter 2 are analyzed in Chapter 6, in relation to both national and international institutions. Primary attention is given to the actions of the Dutch government and the effects of those actions.

Because the approach in the Dutch study was considered to be inadequate in terms of generation of information for national and international decisions, a more adequate framework for analysis is presented in Chapter 7.

Concluding observations and continuing questions are presented in Chapter 8.

3 Water quality management plan for the Dutch portion of the North Sea

The ABC study draws heavily on analyses executed by Rijkswaterstaat (one of the Directorates of the Dutch Ministry of Transport and Public Works) and the Delft Hydraulics, in preparation of the WQMP-North Sea. This chapter explains the background and limitations of that study in order to enable a better understanding of the ABC study and its results.

3.1 BACKGROUND AND OBJECTIVES OF THE WQMP-NORTH SEA STUDY

In April 1984, a decision of the Dutch government was published with respect to the note, Harmonization of North Sea Policy. This note had been formulated by ICONA in 1982 and was subsequently subjected to a procedure of public participation and advice. The note comprised an action programme, including the formulation of a Water Quality Management Plan for the North Sea. This action was formulated in the following terms (Ministerie van Verkeer en Waterstaat, 1982):

> "Building on international conventions and actual national policies, the plan will represent the vision of the Dutch Government on a coherent and strategic approach to preventing and combatting pollution of the North Sea. The plan will provide directives for international negotiations and in addition will indicate management strategies as far as they can be taken in the Netherlands." (p. 33)

The leading ministry for drafting the WQMP-North Sea was the Ministry of Transport and Public Works. The study itself was carried out under responsibility of the Directorate North Sea of Rijkswaterstaat. Supporting analyses were performed by specialists of the Directorate North Sea and the Delta Department of Rijkswaterstaat, and the Delft Hydraulics, in an eighteen month period in 1983/1984. The study resulted in a draft WQMP-North Sea, which was presented to the public at the end of 1985.

The focus of the analysis to develop the WQMP-North Sea is on water quality management, in terms of analyzing the effects on water quality

--- in the water column and in sediments --- of the production of a given mix of products and services (levels of activities), for present conditions and for some set of modified conditions resulting from the implementation of steps actually undertaken to reduce cadmium discharges in the Netherlands.

Before providing a short description of the analysis conditions for the WQMP-North Sea supporting studies, it is illustrative to state the general goal of Dutch North Sea water quality management, as expressed in the final version of the note, Harmonization of North Sea Policy (Ministerie van Verkeer en Waterstaat, 1984), i.e.:

> "To maintain or to obtain a water quality in the North Sea which enables preservation of the ecological values, thereby taking into account societal demands for the production of goods and services from the North Sea." (p. 10)

The "preservation of ecological" values is specified as being comprised of:

> "Maintaining or recuperation of a diversity of organisms and of aquatic ecosystems as close as possible to an undisturbed situation" (Ibid, p. 10); and
>
> "Preventing that, due to human activities, changes of the North Sea ecosystem will occur, which are irreversible and have adverse effects on the functions of the North Sea" (Ibid, p. 10).

3.2 CHARACTERISTICS OF THE ANALYSES

The relatively short period for the study, the difficulties involved in obtaining the proper information, and the non-specific objectives for the WQMP-North Sea cited above, implied some major constraints for the possibilities to perform an analysis such as described in Chapter 7 and Appendix A. The constraint of total duration of 18 months was dominant and resulted in an elimination of parts of the projected analysis, e.g., a spatially differentiated analysis of ecologic characteristics to provide some basis for a comparison of relative worth among geographic areas and concentrations of toxic substances.

The following segments of analysis were included: setting up the analysis; estimating discharges from various sources; and analyzing natural systems, including the effects or potential effects on species. Those steps can be compared with those described in Appendix A.

In addition, an investigation was made of international agreements, national regulations, and policy guidelines reflecting the legal and institutional structure for water quality management in the Dutch portion of the North Sea. This information was used in two ways. One, constraints were looked for, in the sense that the legal and institutional structures could not provide the proper instruments to stop deterioration or to obtain an improvement of the actual water quality. Second, formal policy guidelines related to activities using North Sea marine resources, were evaluated with respect to consistency, in particular in relation to the water quality effects of the activities involved.

The characteristics of the output of the study were implicitly defined to be: (1) estimates of discharges from various sources; (2) estimates of concentrations of various substances in water; (3) estimates of effects of those concentrations on species; (4) description of the ecology of the North Sea; and (5) delineation of the legal and institutional context of water quality management for the Dutch portion of the North Sea.

In summary, the following general observations merit emphasis with respect to a characterization of the Dutch study.

- The study concentrated on water quality management and considered mainly those human activities which influence ambient water and sediment quality of the North Sea: shipping; gas and oil mining; and disposal of wastes.
- Alternative mixes of products and services were not analyzed. Rather, estimates were made of inputs of wastes into the North Sea in a kind of present situation, characterized by the year 1980. Developments subsequent to 1980 were assessed only with respect to: (1) the cadmium concentrations in the Rhine in 1985; and (2) the decrease of cadmium discharges by Dutch controlled sources, as a result of concrete actions taken between 1980 and 1985.
- No specific measures to reduce discharges were formulated and, as a consequence, no economic analysis was carried out. For the same reason no use has been made of the collected legal and institutional information to carry out an analysis of how concrete measures might be implemented.
- No analysis was made of the benefits of improved water quality, e.g., in relation to specific ecologic characteristics and specific uses of Dutch coastal waters. This would require a spatial differentiation of such characteristics and uses and of the concentrations of different substances. Part of this information was made available, but time and budget constraints impeded a proper approach in this direction.
- No time was available to carry out sensitivity analyses with respect to the major assumptions, including, e.g.: estimates of atmospheric deposition; the North Atlantic Ocean concentrations applied to the inflow through the English Channel; constant river inflows throughout the year; utilization of tidal averaged flows; and effect of stratification due to temperature and salinity differences in the water column.

More specific analysis conditions relate to the area, substances, and sources considered. The study or planning area coincided with the Dutch portion of the North Sea as shown in Figure 1-2. For the purpose of water circulation and water quality modelling, the horizontal boundaries coincided with the so-called southern North Sea, bounded on the north by the parallel of 56 degrees latitude and in the south by the line across the English Channel, both as shown in Figure 1-2.

Substances considered were: cadmium (Cd); mercury (Hg); lead (Pb);

copper (Cu); chromium (Cr); zinc (Zn); phosphorus (P); and nitrogen (N). Although these substances occur in both dissolved and particulate forms, only total concentrations were considered. As part of the description of the ecology of the North Sea, the toxic effects of organochlorides were investigated. An estimate was made of discharges of petroleum hydrocarbons (oil) into the North Sea. Based on this analysis, oil was not considered to be a major problem, as of 1984.

Sources considered were: atmospheric deposition; riverine inputs; oceanic cross boundary flows; coastal outfalls; dumping of dredging spoils; dumping of industrial wastes; dumping of sewage sludge; incineration on ships; marine transport; and offshore oil and gas operations.

3.3 EVALUATION OF AMBIENT WATER QUALITY

The Dutch study in preparation of the WQMP-North Sea developed a methodology to establish critical and target concentrations in the North Sea, in order to decide on measures, if any, to be taken. An attempt was made to relate target concentrations for various substances to effects of those substances in the marine environment.

Referring to the effects of actual concentrations was not possible, because there were no clear indications that some of the symptoms of the effects of toxic substances in the North Sea on organisms were caused by certain ambient concentrations --- in water and in sediments --- of considered substances. Therefore, target concentrations were based on the results of toxicologic laboratory tests.

Effects can be divided into lethal effects, causing an organism's death, and sublethal effects, comprising all noticeable effects not leading to death. Laboratory tests can be carried out to estimate lethal and sublethal concentrations, i.e., concentrations in water causing lethal and sublethal effects, respectively, and the no-effect level, representing the highest concentration which does not cause noticeable effects.

Lethal concentrations (LC), sublethal concentrations (SLC) and no-effect levels (NEL) were developed for particular substances in relation to: specific organisms; a specified time span, 96 hours unless indicated differently; and a certain percentage of organisms on which the effects were observed, 50% unless indicated differently. For example, for a specific organism a LC of cadmium of 5 mg/m^3 means that, in laboratory tests 50% of this organism died within 96 hours in an aquarium with water containing 5 mg/m^3 of cadmium.

Using such results for the analysis of measured or estimated concentrations in the North Sea includes at least the following uncertainties.

- As already observed in Chapter 2, in the real world organisms react to a combination of different substances, referred to as synergetic and antagonistic effects.
- Toxic effects of heavy metals may appear only after a long period, due to their bioaccumulation characteristics.

- Organisms are exposed to toxic substances not only through water but through the food they consume.
- Toxicological experiments often deal with dissolved concentrations in water, in which form substances are biologically available. In the WQMP-North Sea study only total concentrations were used. However, not all substances in the North Sea are totally dissolved. For example, a substantial part of heavy metals is adsorbed to suspended material and sediments.

Lethal concentrations were used in the WQMP-North Sea to develop target concentrations associated with factors of safety to compare with the estimated concentrations in the North Sea. They were also used with a second indicator, the anthropogenic fraction, to establish priorities among different toxic substances with respect to the allocation of available budget and administrative operational capacities. The anthropogenic fraction is the fraction of the concentration at any location which was estimated to result from human activities.

Lethal and target concentrations for cadmium

Based on a review of available literature on toxicologic experiments, the following lowest concentrations were found for exposure to cadmium of phytoplankton, zooplankton, benthic organisms, and fish: NEL <1 mg/m^3; SLC: 1 mg/m^3; and LC: 5 mg/m^3. As mentioned above, various uncertainties are attached to the results of laboratory tests, particularly when extrapolating the results to field conditions. Therefore, in defining a target level, i.e., a desired limiting concentration not to be exceeded because of presumed possible adverse effects, a factor of safety was applied to the lethal concentrations found. Applying a factor of safety of 100 to the lethal concentration found for cadmium, 5 mg/m^3, yields a target concentration of 0.05 mg/m^3, i.e., 5 mg/m^3 divided by 100. It should be noted that this target level is 1/20th of the NEL of about 1 mg/m^3.

The factor of safety of 100 is consistent with the factor of safety used by the U.S. Environmental Protection Agency (Train, 1979). However, the Environmental Protection Agency (EPA), using this factor of safety, arrived at an ambient standard for cadmium of 5 mg/m^3. This implies that the EPA calculated this standard using a lethal concentration of 500 mg/m^3, which is 100 times the lethal concentration found in the literature, as noted above. This difference illustrates another problem related to the use of results of laboratory tests, namely the fact that aquatic toxicology is still in its infancy. There is no generally accepted information on lethal concentrations of dangerous substances such as cadmium.

Figure 3-1 shows areas of the southern North Sea where cadmium concentrations were estimated to exceed 0.05 mg/m^3 for 1980 winter conditions.

3.4 CONCLUSIONS IN THE WQMP-NORTH SEA WITH RESPECT TO CADMIUM

Despite the many uncertainties inherent in the analysis, the results

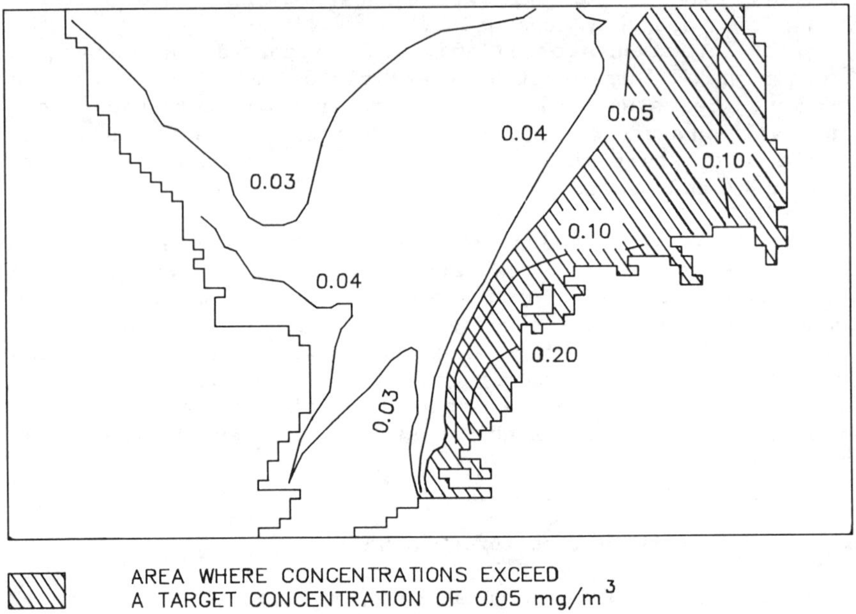

Figure 3-1. Areas in the Southern North Sea where for 1980 winter conditions a target concentration of 0.05 mg/m^3 is estimated to be exceeded

shown in Figure 3-1 were sufficient for the Dutch government to decide to take additional measures to reduce the input of cadmium to the North Sea. In the opinion of the Dutch government, the effects of cadmium --- particularly in the long run --- are unknown and cadmium discharges to the environment from human activities should be eliminated as much as possible.

In the executive programme 1986 - 1989 of the WQMP-North Sea, the following statements and recommendations refer to cadmium (Rijkswaterstaat, 1985).

- Further improvement of water quality must be pursued, in addition to the improvements which will be achieved by current and planned actions.
- With a view to reduction of North Sea pollution, within the framework of the Dutch Surface Water Pollution Act, the elimination of large industrial discharges into estuaries and river mouths is to be effectuated as soon as possible. This will mainly involve those industrial discharges which are expected to contain considerable amounts of cadmium in 1990.
- Research should be intensified with respect to: atmospheric deposition; non-point sources; and the contribution of national discharges and transnational loads of cadmium to the North Sea.

4 Estimating discharges under 1980 conditions

4.1 INTRODUCTION

In this chapter estimates of discharges are given for what was called the present situation at the time data were collected for the WQMP-North Sea. This situation is characterized by the year 1980, but does not represent specific 1980 economic and hydrologic conditions for all discharges. More detail is given in subsequent sections. The area under consideration is the southern part of the North Sea, which is the part south of the 56 degree north latitude. This area is shown in Figure 4-1. Special attention was given to cadmium inputs into the Dutch portion of the North Sea, which area is also shown in Figure 4-1.

Inputs from the following sources were considered:

- atmospheric deposition;
- riverine inputs;
- oceanic cross boundary flows;
- coastal outfalls;
- dumping of dredging spoil;
- dumping of industrial wastes from ships;
- dumping of sewage sludge from ships;
- incineration of industrial wastes;
- marine transport; and
- offshore oil and gas operations.

Incidental discharges, e.g., due to shipping accidents or blow outs from oil exploitation operations, were not considered. Another limitation is that only total, dissolved plus particulate, concentrations were considered. No attempt was made to differentiate between the two forms, despite the difference between them in biological availability.

The methodology applied in estimating cadmium inputs from each of the above mentioned sources is discussed in the first part of this chapter. The second section gives the results of these estimates. A description of the main Dutch controlled sources of cadmium is given in the last section.

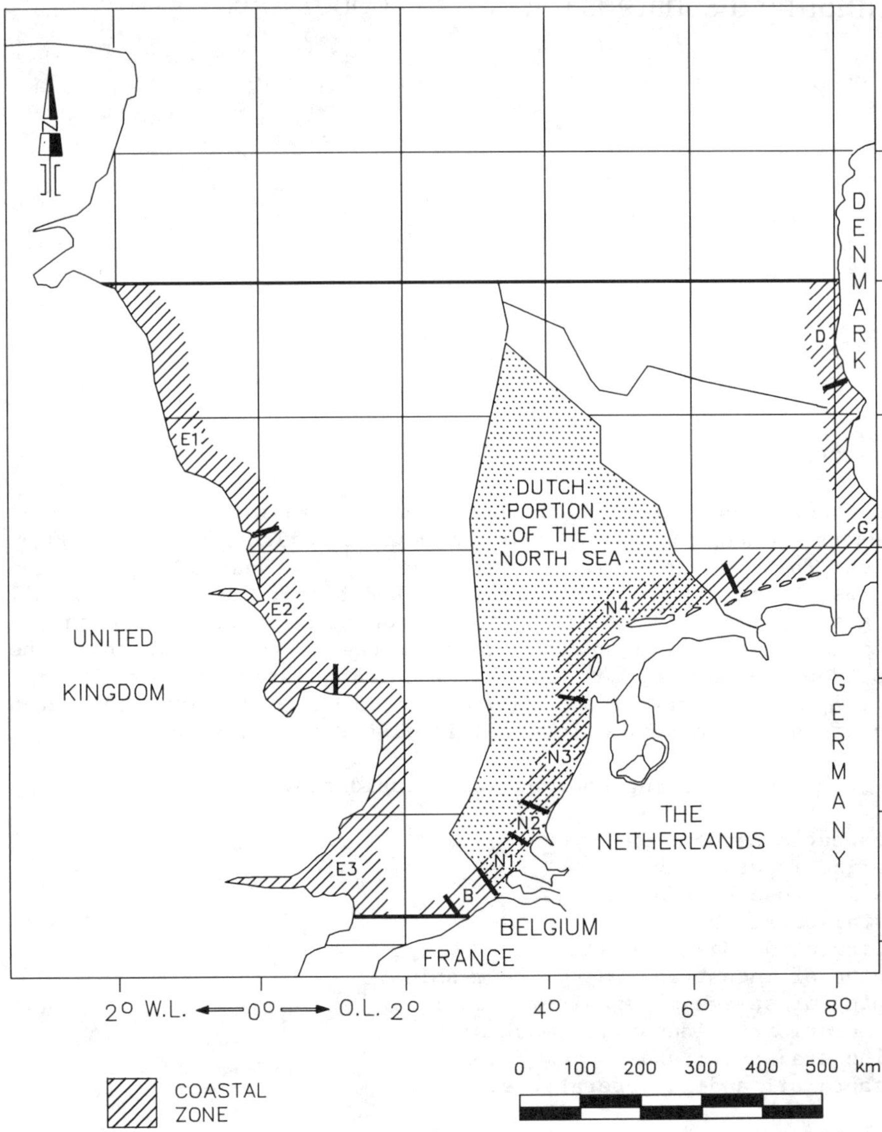

Figure 4-1. Area of North Sea under consideration

4.2 METHODOLOGIES FOR ESTIMATING DISCHARGES

Atmospheric deposition

No observations of deposition rates in kilograms per square kilometer per year ($kg/km^2/yr$) on the North Sea are available. Observations on land sites close to the North Sea and related estimates of yearly deposition rates were available for a period between 1979 and 1982. These estimates show great variation (Van Aalst, 1983a; Van Aalst, 1983b; Kooy and Stokman, 1983). Based on these estimates, low, average, and high values could be calculated. The deposition rate for cadmium used in the WQMP-North Sea study represents the average value, namely 0.3 $kg/km^2/yr$. In the computation of total atmospheric deposition, this rate was applied to the whole area under consideration; no differentiation could be made in deposition rates as a function of the distance to the shores or to specific sources of emissions.

Riverine inputs

Estimates of discharges of cadmium wastes into the North Sea via non-Dutch rivers were taken mainly from reports to the North Sea conference in Hamburg in 1984 (Anonymous, 1984), mainly representing 1980 conditions. For Dutch rivers available water quality plans and reports on water quality were used (Dijkzeul, 1981; Dijkzeul, 1982; Instituut voor Hygiene en Epidemiologie, 1983; International Water Tribunal, 1983; Kooy, 1982; Kooy and Stokman, 1983; Samenwerkende Rijn- en Maaswaterleidingbedrijven, 1983; Rijksinstituut voor Zuivering van Afvalwater, 1983; Rijkswaterstaat, 1983; Salomons, et al., 1981; Wit, et al., 1982). An extensive monitoring network exists in the Netherlands. Data on water levels, flow rates, and water quality are collected at least once a month at various places in internal and coastal waters.

For the cadmium input by the river Rhine, a major source, a special procedure was followed. 1980 data for the Rhine river were corrected to reflect average flow conditions. The correction of - 20% reflecting 1980 as an above average flow year, was based on a rough relationship found between total sediment loads and river flows over the period 1970-1982, i.e., an increasing load for increasing river flows. This may be due to higher erosion rates upstream, or lower retention rates of sediments in estuaries, rivers and harbour basins in years with higher than average river runoff. The reverse may occur for river flows below average.

For inputs by other rivers, which often encompass discharges into estuaries and tidal waters, retention of substances due to, e.g., sedimentation, was not taken into account. For this reason the discharges of cadmium in such cases may represent upper limits.

Part of the cadmium discharge by rivers is not directly related to anthropogenic sources. This so-called natural load for cadmium in solution has been deduced from current concentrations in rivers without direct discharges of human activities and current concentrations in lakes without such discharges and with long residence times. For cadmium load adsorbed on suspended material, erosion rates for the upstream Rhine river basin were used, combined with adsorption rates on suspended material as measured on Rhine sediments in 1900 (Waterloopkundig Laboratorium, 1984). Corresponding concentrations are 0.05 mg/m^3 for cadmium in solution and 0.02 mg/m^3 for cadmium adsorbed to suspended material. This last figure is composed of a suspended material concentration of 20 g/m^3 and an absorption rate of 0.001 mg/g dry weight.

Oceanic cross boundary flows

Estimates of north bound residual flows through the English Channel vary strongly (Otto, 1983). Model computations based on available data on sea currents in the period between 1949 and 1972 (Prandle, 1978) resulted in monthly values showing an overall average of 150,000 m^3/s. Such residual flows are induced both by tide and wind effects, each accounting, as an average, for about 50% of the total residual flow. As wind effects are considerably stronger in winter than in summer, estimates were made of the seasonal flows, using existing models from the Dutch Meteorological Institute. These calculations resulted in the following values for average English Channel inflow under summer and winter conditions:

summer: 75,000 (tide) + 25,000 (wind) = 100,000 m^3/s; and
winter: 75,000 (tide) + 125,000 (wind) = 200,000 m^3/s.

Incoming flows through the North Atlantic Ocean boundary are estimated from model calculations described in Chapter 5, once the water levels and the English Channel inflow are set.

Dissolved concentrations of cadmium in inflowing water masses from the English Channel and the North Atlantic Ocean were assumed to have values found in open ocean waters. For the particulate concentration an adsorption rate of 0.0003 mg/g was used, in combination with measured suspended material concentrations at the English Channel and North Atlantic boundary of 5 g/m^3 and 1 g/m^3, respectively (v.Eck et al., 1983). Summarizing values are presented in Table 4-1.

	Channel	North Atlantic Ocean
Cd in solution (mg/m^3)	0.025	0.025
Cd adsorbed to suspended solids (mg/g)	0.003	0.003
concentration suspended solids (g/m^3)	5	1
total natural Cd concentration (mg/m^3)	0.027	0.025

Table 4-1. Natural concentrations applied for English Channel and North Atlantic Ocean water

In this estimation, the assumption was made --- mainly due to lack of information --- that no cadmium discharged by French rivers enters the southern North Sea through the residual northward bound sea currents in front of the French coast. This is, however, a dubious assumption.

Coastal outfalls

Coastal outfalls represent discharges by industrial activities and municipal sewer systems to the North Sea and to estuaries and rivers under tidal influence, flowing into the North Sea area under consideration.

Data have been taken from the same reports as the data on river inflows, i.e., reports to the North Sea conference in Hamburg, 1984, and water quality plans and other Dutch documents. The references were cited above under riverine inputs.

Dumping of dredging spoil

Data on amounts of dredging spoils from harbours and navigation channels and corresponding cadmium loads were obtained from reports to the Oslo Commission (Oslo Commission, 1984). About 90% of the net cadmium input from disposal of dredging spoil into the North Sea originates from the Netherlands. For the Dutch situation the average of the yearly loads in 1979, 1980, and 1981 was used. For the United Kingdom and Belgium, 1981 values were used.

The gross discharges had to be reduced because part of the cadmium is from marine origin and should be characterized as "recirculation". The recirculating marine part of the sediment was estimated for each dumping site. To determine the recirculating cadmium loads, the average concentration of 0.001 mg/g dry weight in sediments along the southern Dutch coast was used. Table 4-2 gives an overview of the gross and net inputs for three different dumping sites. The natural riverine part in the cadmium input through dredging spoils is estimated to be about 1 ton, or 3% of the net input.

Dumping of industrial wastes and sewage sludge from ships

Dumped amounts and corresponding cadmium loads were obtained from reports to the Oslo Commission (Oslo Commission, 1984) and to the North Sea conference in Hamburg in 1984 (Anonymous, 1984). Data represent 1981 conditions.

Incineration of industrial wastes

On the Dutch portion of the North Sea, industrial wastes --- mainly halogenated hydrocarbons --- from different countries are burned, which may result in local discharge through atmospheric deposition. However, these amounts are so small that they can be neglected in the quantitative analysis.

Marine transport

Marine transport presents an important source of oil discharges to the North Sea. For cadmium, the contribution of this activity can be neglected.

Dumping location [3)]	Loswal Noord		Scheveningen	IJmuiden
Origin of dredgings	Hoek van Holland	Rotterdam	Scheveningen	IJmuiden
Total amount dumped (in 10^6 m^3)	13.6	3.3	0.2	4.4
Dry weight (in 10^6 tons)	6.5	1.6	0.09	1.5
Marine fraction[1)]	95%	50%	100%	100%
Total cadmium load (in tons)	25.7	17.2	0.24	4.0
Estimated recirculation (in tons)	6.0	0.8	0.1	1.5
Net cadmium input (in tons)	19.7	16.4	0.1	2.5[2]

1) Cadmium concentration 0.001 mg/g dry weight
2) Has been included in river inflow
3) Locations indicated in Figure 4-5

Table 4-2. Gross and net inputs of cadmium in dumped dredging spoil, 1980

Offshore oil and gas operations

Estimates for discharges by oil and gas operations in the North Sea are based on data on drilling and production activities in 1980 from the Dutch Ministry of Economic Affairs (Ministerie van Economische Zaken, 1983) and on assumed average concentrations of substances in waste liquids. The following assumptions were made.

- Average boring depth ranges from 3000 m to 3500 m in the southern North Sea, and from 3500 m to 4000 m in the central and northern part.
- An average value of 0.33 m^3 of drilling mud was assumed to be discharged for each meter drilled. This value was obtained from studies made in relation to oil and gas activities in the Gulf of Mexico (Menzie, 1982; National Oceanic and Atmospheric Administration, 1984; Petrazzuolo, 1981).
- Product water in oil and gas production varies, among other things with the age of the exploited fields. An average value of 20% product water in crude oil extracted was applied for the North Sea. The

total amount of product water in gas produced in the North Sea appears to be negligible.

- Data on cadmium concentrations in waste drilling muds and product water are lacking for the North Sea. Therefore the following data for the Gulf of Mexico were applied (Petrazzuolo, 1981):
  - drilling muds: 1.0 g of cadmium per m^3; and
  - product water: 0.34 g cadmium per m^3.

For the total southern North Sea, cadmium input as a result of offshore oil and gas operations was estimated to be about 0.5 tons in 1980.

4.3 RESULTS OF ESTIMATES

The results of the preceding estimates presented in the third background document to the WQMP-North Sea (Rijkswaterstaat and Waterloopkundig Laboratorium, 1985c) are summarized in Table 4-3, Figure 4-2, Table 4-4, and Figure 4-3. Table 4-3 shows the estimated cadmium inputs into the Southern North Sea in 1980, in total about 280 tons, excluding oceanic cross boundary inflow, i.e., from both North Atlantic Ocean and English Channel. Figure 4-2 shows the estimated fractional contributions from various sources to the total cadmium inputs. Table 4-4 shows the estimated cadmium inputs into the Dutch portion of the North Sea in 1980, in total about 150 tons. Figure 4-3 shows the estimated fractional contributions of Dutch sources and of inputs into the Dutch portion of the North Sea to the total cadmium inputs into the southern North Sea.

The major sources for cadmium inputs are the rivers, accounting for about 50% of the input into the southern North Sea. Roughly 2/3 of the riverine inputs comes from the Rhine, thus providing about 70% of the total cadmium input into the Dutch portion of the North Sea.

Atmospheric deposition is highly uncertain. Best estimates were 25% of the total input for the southern North Sea and 10% for the Dutch portion. The major part of the atmospheric deposition of cadmium in the North Sea derives from cadmium emissions into the atmosphere by four activities: coal combustion; iron and steel production; non-ferrous metallurgic industries; and waste incineration. All surrounding countries, i.e., Belgium/Luxemburg, Denmark, United Kingdom, France, Germany (FRG), and the Netherlands, contribute to cadmium emissions into the atmosphere.

For the southern North Sea the coastal outfalls, mainly those in the United Kingdom, and dumping of dredging spoil, mainly in front of the Dutch coast, are substantial, with 9% and 14%, respectively, of the total cadmium input.

An estimate was made of the cadmium levels in the open Atlantic Ocean. These estimates were then used to estimate the total inflow of cadmium into the southern North Sea through the northern and southern boundaries. As observed previously, this procedure neglects possible inputs from sources south of the English Channel. The cadmium input through the boundaries, under the above assumption, proved to be in the same order of magnitude as the total of the above mentioned discharges from

Location[1)]	Rivers			coastal	dumping from ships			incineration	marine	off-shore	total excl. atmospheric deposition			atmospheric deposition	total	boundary[2)]
	nat.	anthrop.	total	outfalls	dredge spoils	industr.	sewage sludge	chem.wastes	transport	operations	nat.	anthrop.	total			
N1	0.6	15.9	16.5	1.1	0	0.7	0	0	<	0	0.6	17.7	18.3	0.3	18.6	
N2	5.1	61	66.1	0	36.4	0.1	0	0	<	0	5.1	97.5	102.6	0.3	102.9	
N3	0.2	1.5	1.7	1.4	0.1	0	0	0	<	0	0.2	3.0	3.2	0.7	3.9	
N4	1.4	10	11.4	0.1	0	0	0	0	<	0	1.4	10.1	11.5	1.0	12.5	
Total Dutch Coast	7.2	88	96	2.6	36	0.8	0	0	<	0	7.2	129	136	2.3	138	
E1	0.3	5.9	6.2	3.5	3.8	0.3	0	0	<	0	0.3	13.5	13.8	1.3	15.1	
E2	0.7	10.0	10.7	3.3	0	0	0.3	0	<	0	0.7	13.6	14.3	1.7	16.0	
E3	0.5	3.2	3.7	14.6	0	0	4.0	0	<	0	0.5	21.8	22.3	1.1	23.4	
Total UK Coast	1.5	19	21	21	3.8	0.3	4.3	0	<	0	1.5	49	50	4.1	54	
Total Belgian Coast	0.1	0.1	0.2	0	0	0.1	0	0	<	0	0.1	0.1	0.2	0.5	0.7	
Total German Coast	3.7	16.3	20	0	0	0	0.4	0	<	0	3.7	16.7	20.4	1.6	22	
Total Danish Coast	0.2	3.4	3.6	0	0	0	0	0	<	0	0.2	3.4	3.6	0.7	4.3	
Total Coastal Waters (%)	13	130	140	24	40	1.2	4.7	0	<	0	13 (6)	200 (90)	210 (96)	9.2 (4)	220	
Remaining North Sea below 56° NB					0	0	0	0.04	<	0.5	0	0.5	0.5	57	58	280
Total North Sea (%)	13 (5)	130 (46)	140 (51)	24 (9)	40 (14)	1.2 (0)	4.7 (2)	0.04 (0)	< (<)	0.5 (0)	13 (5)	200 (72)	210 (76)	66 (24)	280	280 (100)

1) See Figure 4-1

2) Inflow through the English Channel in the South and from the Atlantic Ocean in the North Sea
values in tons per year
0 = known, non-existing or negligible, < = unknown, but expected to be negligible
nat. = natural fraction of riverine inputs, anthrop. = anthropogenic fraction of riverine inputs
totals are rounded

Table 4-3. Summary of estimated cadmium inputs into the southern North Sea in 1980

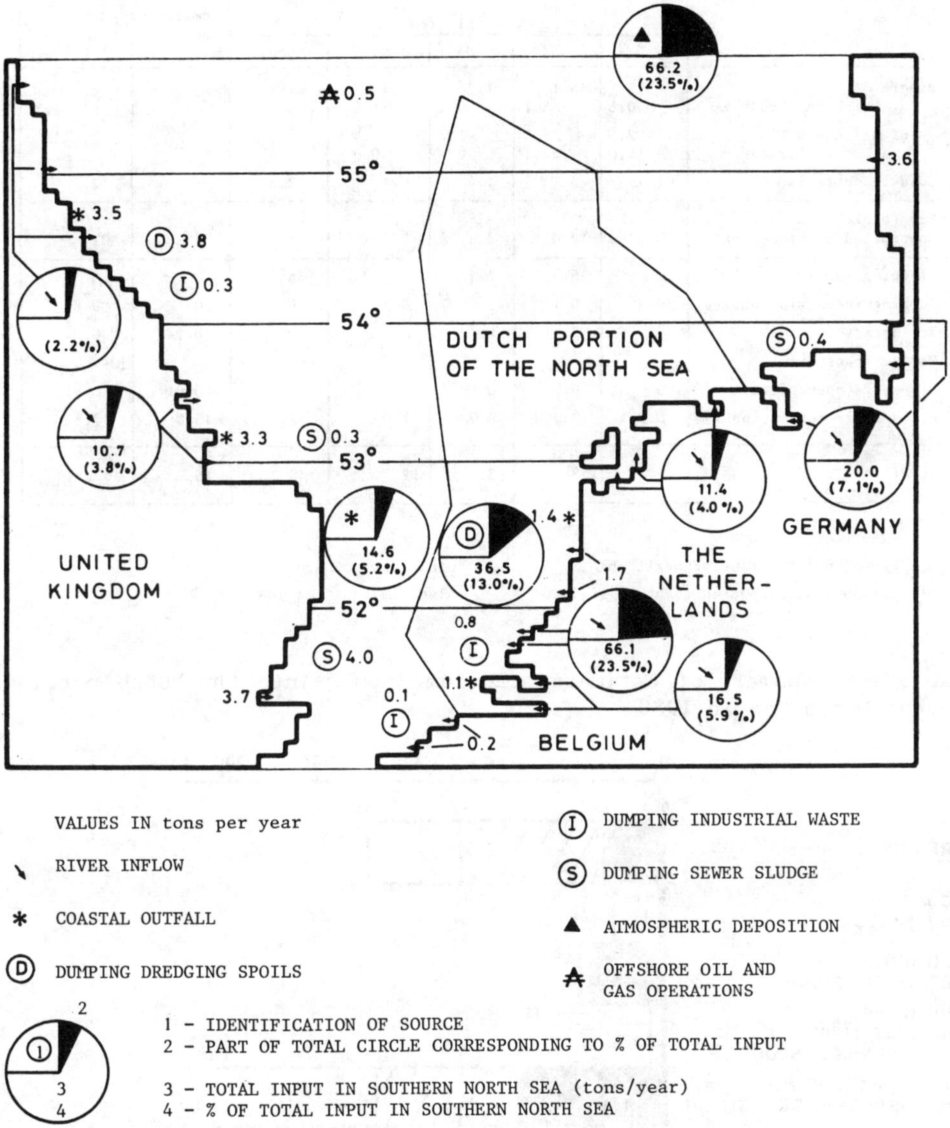

Figure 4-2. Estimated fractional contributions of various sources to total cadmium inputs into the southern North Sea in 1980

the specified sources. This makes it rather critical to check the French contribution.

The total input controlled by the Dutch, excluding atmospheric deposition, is about 70 tons: 32 tons through rivers and coastal outfalls, 37 tons through dumping of dredging spoil, and about 1 ton through dumping of industrial wastes from ships.

	Coastal area[1]					Remaining part	Total
	N1	N2	N3	N4	Total		
Rivers	16.5	66.1	1.7	11.4	96		96
of which "natural"	0.6	5.1	0.2	1.4	7.3		7.3
of which - Dutch	2.5	19.0	1.7	5.4	29		29
- foreign	14.0	47.1	0	6.0	67		67
Coastal outfalls	1.1	0	1.4	0.1	2.6		2.6
Sub total Dutch controlled contribution	3.6	19.0	3.1	5.5	31		31
Dredging spoil	0	36	0.1	0	36	0	36
Dumping industrial wastes	0.7	0.1	0	0	0.8	0	0.8
Incineration	0	0	0	0	0	0.04	0.04
Marine transport	<	<	<	<	<	<	<
Off-shore operations	0	0	0	0	0	0.01	0.01
Atmospheric deposition	0.3	0.3	0.7	1.0	2.3	15.0	17.3
Total	19	103	3.9	12	138	15	153

[1] See Figure 4-1

0 = known; non-existing or negligible — totals are rounded

< = unknown, but expected to be negligible — values in tons per year

Table 4-4. Summary of estimated cadmium inputs into the Dutch portion of the North Sea in 1980

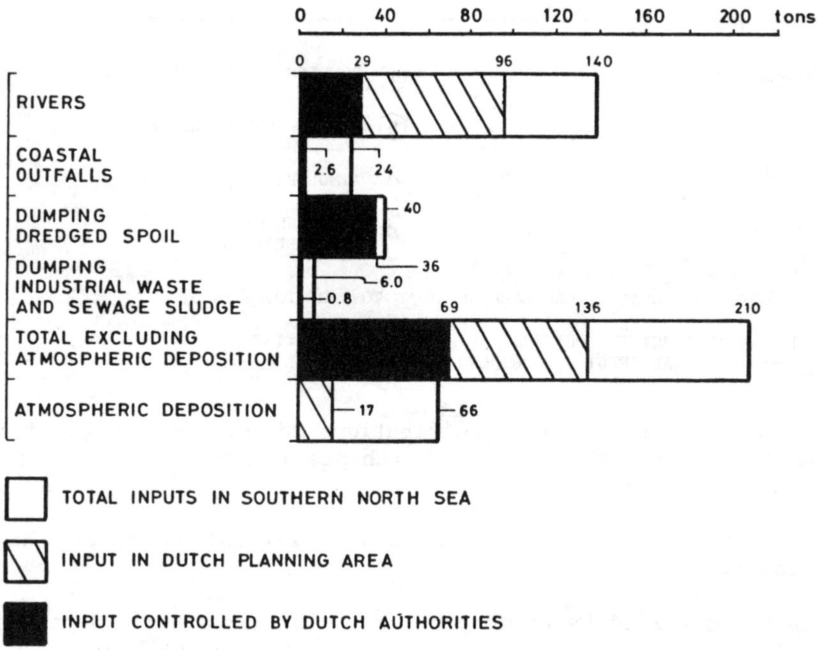

Figure 4-3. Estimated cadmium inputs in 1980 of various sources into the southern North Sea and the Dutch planning area, differentiating inputs controlled by Dutch authorities

4.4 DUTCH CONTROLLED CADMIUM SOURCES

In this section a short description is presented of cadmium discharges, in or about the year 1980, which are or can be controlled by Dutch authorities. Where possible an estimation is given of the contribution of such discharges to the cadmium inputs into the North Sea. This section is intended to give a more detailed impression than in the previous section of the cadmium sources in the Netherlands. As such, it indicates: (1) the Dutch contribution to the total cadmium load of the North Sea; and (2) the possible measures in the Netherlands for reducing this contribution. Though intended, it proved to be impossible to obtain similar figures for the situation in the United Kingdom within the limited time and budget available for this study.

The amounts of cadmium discharged in or about 1980 are presented in Figure 4-4. Before giving more information on Dutch sources of cadmium, a few general remarks are presented with respect to the pathways of cadmium shown in Figure 4-4.

The Dutch part of riverine cadmium inputs into the North Sea derives from discharges of waste water treatment plants (WWTPs), urban storm runoff, industrial activities, and agricultural operations. Part of the cadmium in discharges and in surface waters will be dissolved, part will be adsorbed to suspended material and sediments. The dissolved fraction flows out into the North Sea. The adsorbed fraction may accumulate in riverine and estuarine sediments. However, the downstream areas of main Dutch rivers are dredged regularly. Some portions of the dredging spoils, containing cadmium, are dumped into the North Sea. Thus, by riverine input as well as by dumping dredging spoils, cadmium from Dutch aquatic discharges may end up in the North Sea.

It has to be noted that all known Dutch cadmium discharges into rivers described in this section, in total 23.1 tons, do not account totally for the Dutch part of cadmium inputs by river inflow as indicated in Table 4-4. In the WQMP-North Sea study, the Dutch fraction of riverine cadmium inputs was estimated to be 29 tons. This estimation was based on flow figures of all inflowing rivers and waterways and measured or estimated cadmium concentrations in those flows. Possible explanation(s) of this difference may be one or more of: an underestimation of the contribution of non-point sources to surface waters; the fact that not all cadmium emitting industrial activities are accounted for; and an overestimation of cadmium concentrations in inflowing rivers and waterways.

However, the emerging picture is clear. The fertilizer industry accounts for about 60% of known Dutch controlled aquatic cadmium discharges, and for about 25% of the total Dutch controlled cadmium input into the North Sea, which is 69 tons, including aquatic discharges and dumping operations and excluding atmospheric deposition. The dumping of dredging spoil represents almost 60% of this total input to the North Sea.

Dutch cadmium discharges into the atmosphere were estimated to be 4.1 tons in 1980. There is, however, no accurate information about the relation between Dutch cadmium discharges into the atmosphere and depo-

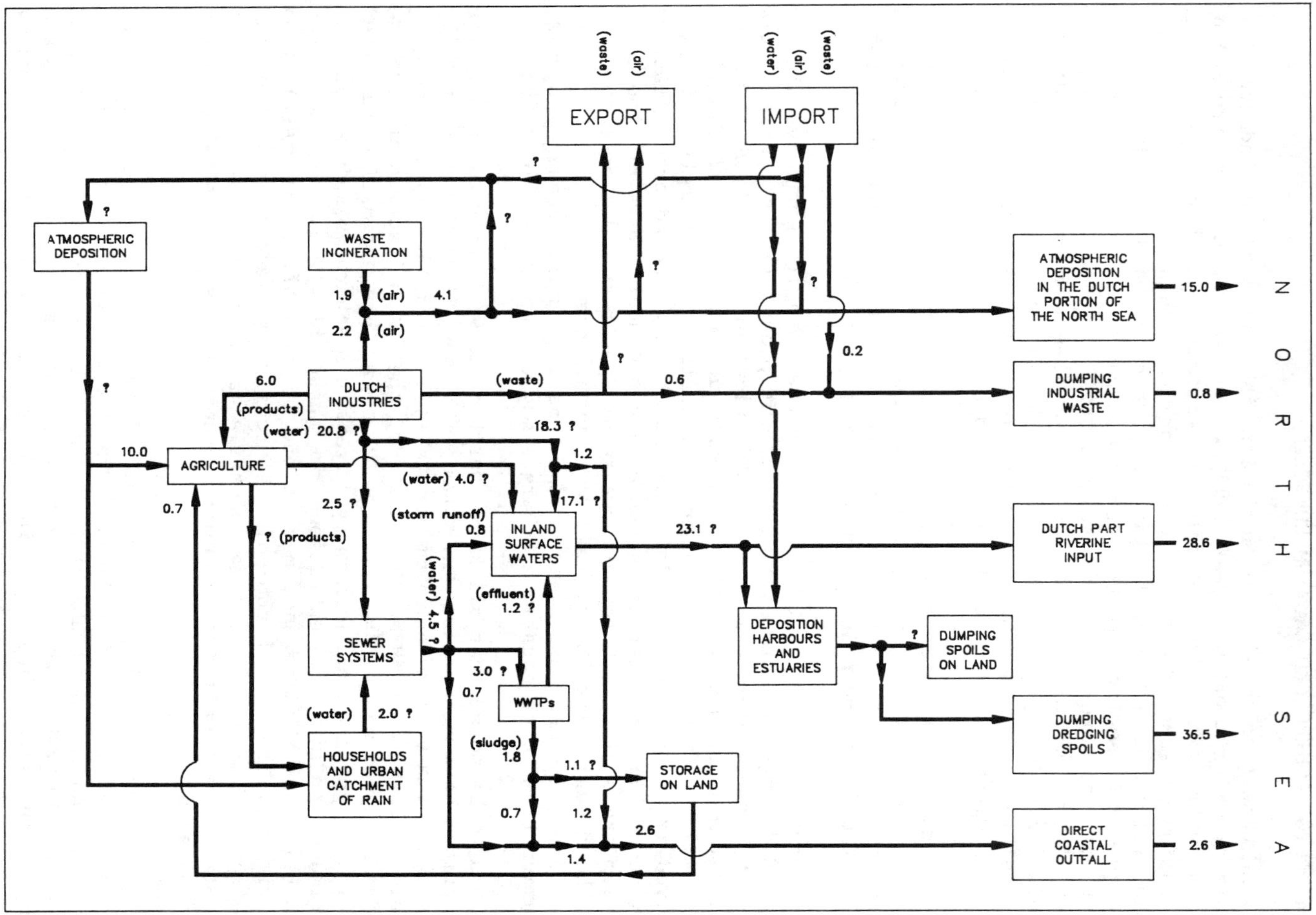

Figure 4-4. Pathways of Dutch controlled estimated cadmium discharges in 1980

sition of those discharges in the North Sea. Given the predominant westernly, northwesternly, and southwesternly winds, much or most of those discharges are probably transported away from the North Sea.

Detailed information on the Dutch cadmium discharging activities on such items as: applied production processes and alternatives; production costs and relevant components; and expected developments, proved to be hard to obtain. Where obtained, they have been included in the subsequent description of Dutch sources of cadmium. The description pays attention to:

- sewer systems and waste water treatment plants (WWTPs);
- industrial activities discharging directly to inland surface waters and estuaries;
- agricultural operations;
- dumping from ships; and
- atmospheric discharges by Dutch industrial activities and waste incineration plants.

Sewer systems and waste water treatment plants

Through municipal sewer systems in 1980, industries discharged about 2.5 tons of cadmium (Vliet and Feenstra, 1982). The most important type of industrial source of cadmium discharges to sewer systems is the electroplating industry. In 1980 these industries discharged 1.5 tons of cadmium (Ministerie van Volkshuisvesting, Ruimtelijke Ordening en Milieubeheer and Ministerie van Verkeer en Waterstaat, 1985).

By domestic discharges and the catchment of rainfall about 2 tons of cadmium ended up in sewer systems. The estimates for these discharges are based on the following assumptions (Waal-Malefijt, 1983): 45 mg Cd/inhabitant/yr is discharged by household food and beverage consumption and the urban catchment of rainfall accounts for about 115 mg Cd/inhabitant/yr, supposing an atmospheric deposition rate of 0.4 $kg/km^2/yr$. One tenth of Dutch households is not connected to a sewer system.

From the total wastewater flows into sewer systems, in 1980 about two thirds were treated in WWTPs before being discharged into surface waters (Oslo Commission and Paris Commission, 1984). The efficiency of WWTPs in removing cadmium is estimated to be about 60%; this percentage of cadmium is removed from the waste water and adsorbed by the sludge. From figures above it can be calculated that in 1980 about 2.7 tons of cadmium, discharged into sewer systems, ended up in surface waters, either in WWTP effluents or in untreated waste waters, e.g., urban storm run-off. Out of this 2.7 tons, 0.7 ton was discharged directly into the North Sea through a 2.5 km long pipeline near The Hague. In the sewage sludge 1.8 ton of cadmium was adsorbed in 1980, of which about 0.7 ton is discharged directly to the North Sea through a second, 10 km long, pipeline near The Hague.

The destination of the rest of the sewage sludge was in 1980: 65% to agricultural uses; 34% to disposal on land (the cadmium input into surface waters as a result of leaching was assumed to be negligible); 0.3% discharged into surface waters; and 1% incinerated (Oslo

Commission and Paris Commission, 1984). In the subsection "Agricultural Operations", attention will be paid to the cadmium content of the surface water runoff from agricultural land.

Industrial activities

In 1980 the following amounts of cadmium were discharged directly into Dutch inland and coastal waters by major industrial activities (Ministerie van Volkshuisvesting, Ruimtelijke Ordening en Milieubeheer and Ministerie van Verkeer en Waterstaat, 1985):

- fertilizer industry, 16.7 tons;
- metallurgic industry, 0.7 ton; and
- chemical industry, 0.9 ton.

All of these discharges take place in the downstream parts of the Dutch waterway system.

Fertilizer industry
From phosphatic rock and sulphuric acid the fertilizer plants (UKF and Windmill), both located alongside the New Waterway, produce phosphoric acid, an important component of fertilizers. The residual waste gypsum is discharged directly into the New Waterway. In 1981 in total 1600 kilotons of rock were processed, 500 kilotons of phosphoric acid were produced, and about 2000 kilotons of gypsum were dumped (Thomas and Olsthoorn, 1984).

Cadmium occurs as a trace contaminant in phosphatic rock. Concentrations depend on the type of rock. In marine rock processed in Dutch fertilizer plants, the average cadmium concentration is 48 parts per million. About 80% of the cadmium is estimated to end up in the fertilizer, 20% in the waste gypsum.

Metallurgical industry
The major metallurgical plant discharging cadmium into surface waters is a large steel mill (Hoogovens), located alongside the North Sea Channel, the entrance to the port of Amsterdam. Through waste waters discharged into this channel, 0.7 ton of cadmium was discharged in 1980. Cadmium in these waste waters derives from the ferro rock processed in Hoogovens.

Chemical industry
The major chemical plant discharging cadmium into surface waters is Hoechst, located in the southern Delta area. From phosphatic rock, 600 kilotons processed in 1980, Hoechst produces, among other things, washing powder. Waste waters containing about 0.9 ton cadmium were discharged into the Western Scheldt.

Agricultural operations

In Table 4-5 the average cadmium load on agricultural land is indicated. Thomas and Olsthoorn (1984) assumed that as a result of this load, in drainage waters about 1.3 g of cadmium per hectare is "discharged" from agricultural areas into inland surface waters. The agricultural area in the Netherlands is 2.5 million hectares. As a result of atmospheric deposition, there will also be a small cadmium discharge from other soils (e.g., from woods). In total the average cadmium discharge from non-urban areas is estimated to be about 4 tons a year.

	crop land (g/ha)	grass land (g/ha)
fertilizers	5.0	1.6
manure	4.9	2.7
sewage sludge	0.7	-
compost	0	-
atmospheric deposition	4.0	4.0

Table 4-5. Cadmium load on agricultural land in the Netherlands

Dumping from ships

Dumping from ships into Dutch territorial waters included in 1980 the following amounts of cadmium:

- in dredging spoils, 36.5 tons; and
- in industrial wastes, 0.8 ton.

Information on dumping of dredging spoils in Dutch territorial waters was already included in Table 4-2. For dredging spoils from the Rijnmond area --- from Rotterdam to Hook of Holland --- a classification was made, based on the geographical location of the spoils, and, related to that, the degree of contamination (Oslo Commission and Paris Commission, 1984). This classification is:

Class 1 - slightly contaminated material (over 95% marine material);
Class 2 - moderately contaminated material (half marine, half river);
Class 3 - contaminated material (over 50% river material); and
Class 4 - local highly contaminated material.

At present, 1985, Class 1 dredging spoils are all dumped at sea. Some of the Class 2 and Class 3 material is also dumped at sea. Class 4 dredging spoils are dumped on land.

Dumping locations offshore are shown in Figure 4-5. The main Dutch dumping site for dredging spoils is called "Loswal Noord". The major industrial waste dumpings occur in dumping site A near the New Waterway and dumping site B near the Western Scheldt. The largest part of industrial waste dumped in those areas originates from German and Belgian titanium dioxide plants, and consists of sulphuric acid contaminated with organic compounds. The cadmium load in these dumpings, however, is very small, about 0.2 ton in 1980. Although the total amount of dumped wastes originating from Dutch industries is relatively

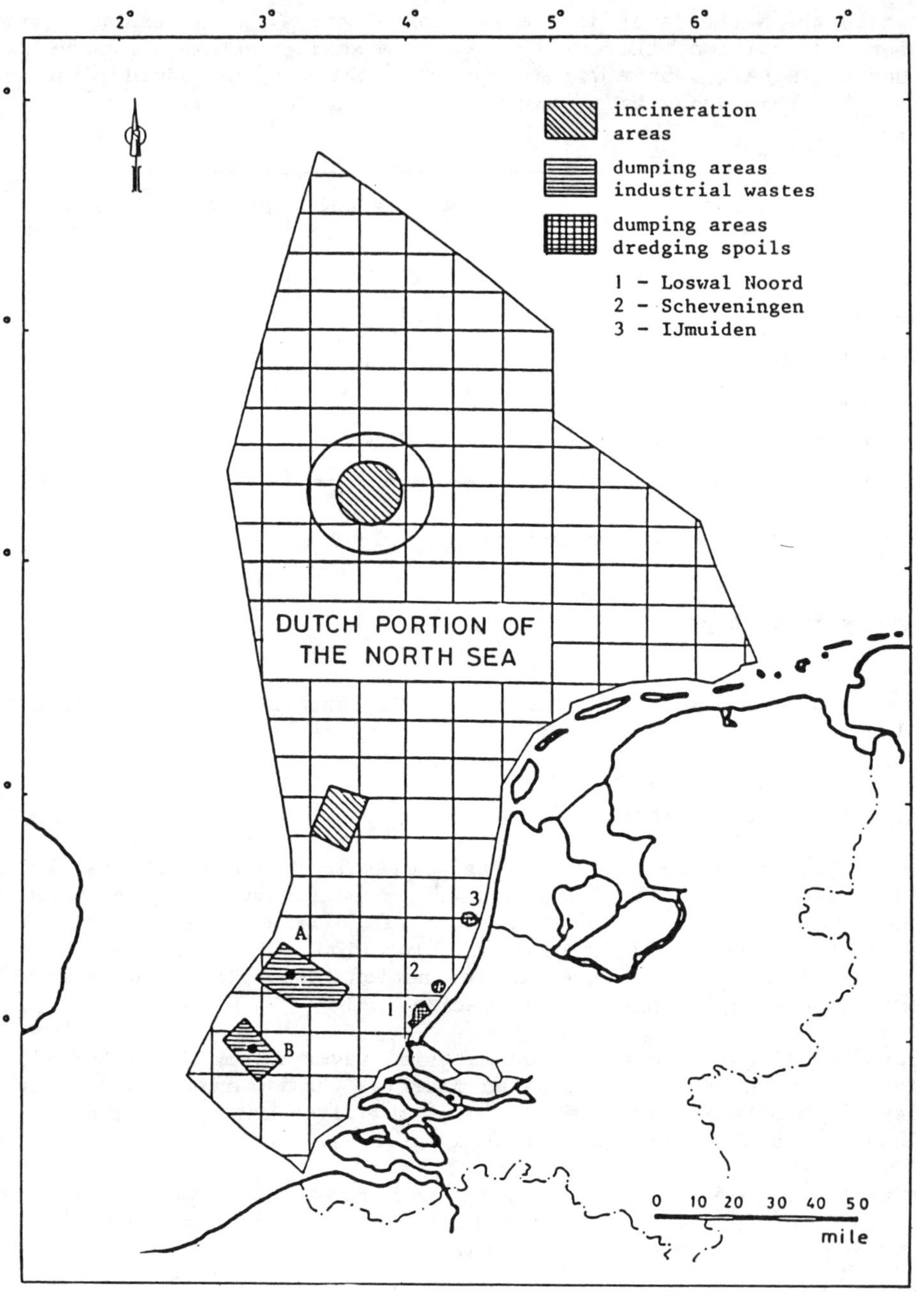

Figure 4-5. Dumping locations in the Dutch portion of the North Sea

small, the cadmium load in these wastes was estimated to be responsible for 75% of the cadmium input into the North Sea by dumping industrial wastes. In 1980 the total cadmium input by dumping industrial wastes was estimated to be 0.8 ton.

Atmospheric discharges

Cadmium discharges into the atmosphere by Dutch industries and waste incineration plants are estimated as follows (Ministerie van Volkshuisvesting, Ruimtelijke Ordening en Milieubeheer, 1986):

- waste incineration, 1.9 tons in 1982;
- metallurgical operations, foundries, 1.2 tons;
- coal combustion, 0.3 tons;
- production of cadmium containing products, 0.3 tons;
- non-ferro metallurgical activities, 0.2 tons; and
- other industrial activities, 0.2 tons.

Major doubts exist in relation to the contribution of the waste incineration plants. According to Thomas and Olsthoorn (1984), the cadmium discharge into the atmosphere by waste incineration plants is substantially higher, 4.5 tons in 1981. However, as pointed out previously with respect to dominant winds, little if any of these Dutch discharges into the atmosphere is likely to reach the North Sea.

5 Estimating effects of discharges on ambient water quality

A marine ecosystem represents a complex set of physical, physicochemical, chemical, and biological processes. Involved are tides, density currents, winds, offshore currents, long distance wave-action, storm events, various species habitats at various depths, various resident and transient species, inputs of wastes from the activities of man, and various atmospheric- oceanic interactions.

Analyzing marine ecosystems is particularly difficult because of: the inflows of materials in the atmosphere above a region from onshore terrestrial ecosystems, with subsequent deposition in coastal and oceanic waters in the region of interest; and the interrelations among tides, long- distance wave action, geological structure, horizontal surface and subsurface currents, and vertical fluid flows, e.g., from deep waters to surface waters and the reverse. Except for relatively enclosed basins beneath the surface in offshore regions, marine ecosystems are relatively unbounded and subject to many more cross-boundary influences than are on- shore terrestrial and aquatic ecosystems. A further difficulty stems from having to distinguish the impacts of man's activities from the impacts of endogenous natural processes, e.g., natural hydrocarbon leaks on the ocean floor, upwelling of concentrated plumes of chemicals. The processes listed above affect, and are affected by, the mix of activities within a marine resources management region as products and services are produced, and by natural processes originating external to the region. Figure 5-1 attempts to depict some of these relationships, as they affect the state of the system (SOS) over time.

Toxic materials, such as cadmium, enter a marine resources region as a consequence of both land-based and offshore based human activities. After discharge to the natural environment, the wastes are affected by, and in turn may affect, various physical, physicochemical, chemical, and biological processes. These processes transform the time and spatial patterns of entering waste materials into the time and spatial patterns of ambient water quality.

This chapter describes how the estimates of ambient water quality and of sediment quality, resulting from discharges of cadmium in 1980, were made. Also presented are the results of the application of the methods for making the estimates, and some data on measured concentrations of cadmium.

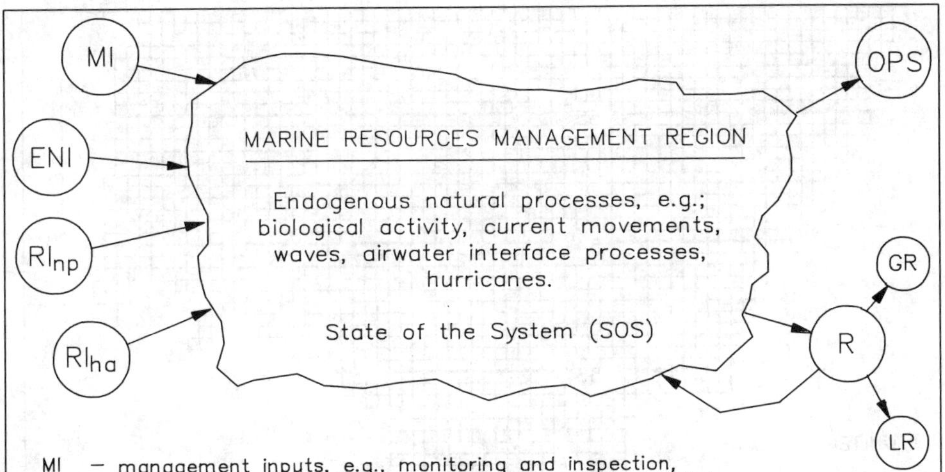

MI — management inputs, e.g., monitoring and inspection, placing of navigation aids, constructing artificial habitat

ENI — endogenous natural inputs, e.g., precipitation, sunlight, wind, temperature

RI_{np} — residuals inputs from outside the region, via natural processes, e.g., current movements, dry and wet deposition from the atmosphere

RI_{ha} — residuals inputs from outside the region, via human activities e.g., sludge transported to and discharged in the region

OPS — outputs of products and services

R — residuals

GR — gaseous residuals discharged beyond the region

LR — liquid residuals discharged beyond the region

SOS at time $t_2 = f$ [state of system at t_1, MI, ENI, RI, OPS, R–GR–LR]

Figure 5-1. Illustration of "State of System" concept in relation to a marine resources management region

5.1 ESTIMATING EFFECTS ON WATER QUALITY

In the Dutch WQMP-North Sea study, a two-part model --- ESTFLO-DELWAQ (Gerritsen, 1983; Postma, 1984) --- was developed to estimate concentrations of conservative materials in North Sea water (Rijkswaterstaat and Waterloopkundig Laboratorium, 1985b). ESTFLO is the hydrodynamic model; DELWAQ is the water quality model. In applying these models, the North Sea south of the 56th degree parallel, north latitude, was divided into the grid system shown in Figure 5-2.

Major assumptions and simplifications

In developing the two-part model, some assumptions and simplifications had to be made with respect to the functioning of the marine ecosystem of the North Sea because of: lack of knowledge; lack of data; and insufficient time and resources for the analysis. Ideally, the effect of these assumptions and simplifications on the results of the analyses should be subjected to sensitivity analysis; this has not been done.

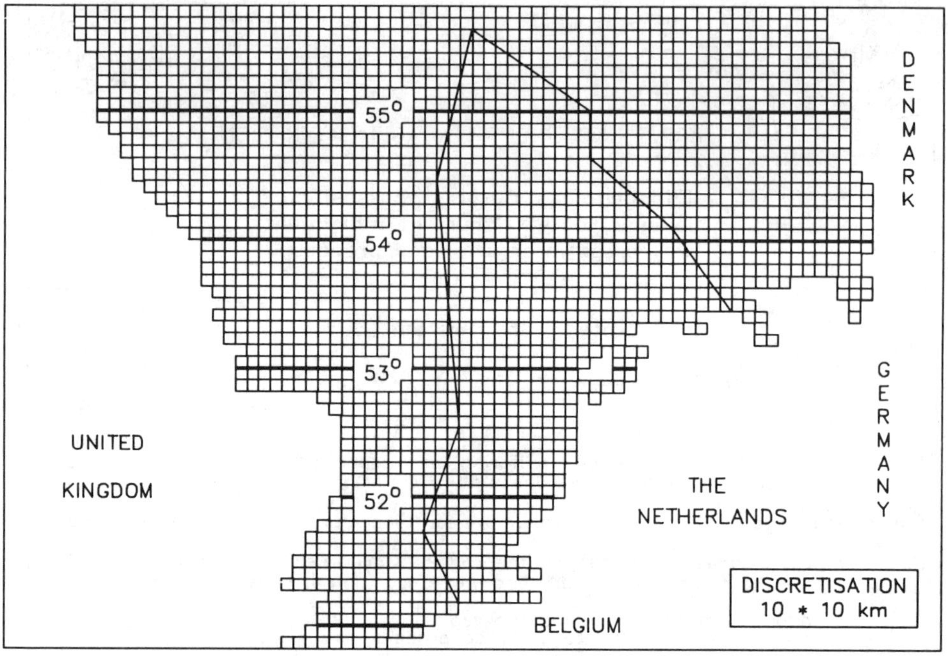

Figure 5-2. Grid system for the North Sea south of 56 degree north latitude

The major assumptions with respect to the water circulation and water quality analyses follow.

- Water circulation modelling differentiated between winter and summer conditions as far as residual currents in the North Sea were concerned. Characteristic winter and summer currents were simulated with respect to wind conditions. Vectorial averaged wind conditions in January 1979 and August 1979 were taken to be representative for winter and summer conditions, respectively.
- No seasonal differentiation was applied for river flows, nor for discharges.
- Calculations for above mentioned average conditions were made for a tidal period of 24 hours and 50 minutes.
- Estimates of cadmium inputs were based on a number of assumptions already discussed in Chapter 4.
- Only total concentrations of considered substances were taken into account.
- Cadmium --- and other materials analyzed --- were considered to be conservative materials in the marine ecosystem. That is, processes such as adsorption and sedimentation, which would affect concentrations in the water, were generally not considered.
- The possible effects of stratification because of temperature and salinity differences were not considered.

Model application

Based on the residual flows calculated by use of the hydrodynamic model ESTFLO, the mass transport and water quality model, DELWAQ, was used to calculate: the spatial distribution of cadmium concentrations in water; the relative contributions of the different sources to the spatial distribution of the water masses and the spatial distribution of cadmium concentrations; and the retention times of the water masses.

The DELWAQ model solves the advection-dispersion equation with reaction terms and waste loads for specified constituents in the water phase. Thus:

$$\frac{dC}{dt} = \frac{d}{dx}\left(D_x \frac{dC}{dx} - U_x C\right) + \frac{d}{dy}\left(D_y \frac{dC}{dy} - U_y C\right) + \frac{d}{dz}\left(D_z \frac{dC}{dz} - U_z C\right) + \text{reactions} + \text{waste loads} \qquad (5\text{-}1)$$

where C = local concentration;
$D_{x,y,z}$ = dispersion coefficients in x, y and z directions;
$U_{x,y,z}$ = velocities in x, y and z directions; and
t = time.

Dispersion coefficients have been calibrated by a comparison between measured (International Council for the Exploration of the Sea, 1979; Kooy and Stokman, 1983; Suylen, 1981) and simulated salinity distributions. A detailed calibration was not possible, because data on the measured salinity distribution were not available for the whole area considered. By using a dispersion coefficient of 150 m^2/s, the simulated salinity distribution in front of the Dutch coast corresponded best with the measured salinity distribution in that area. The reaction terms represent changes in concentration due to other physical, chemical and biological processes, including exchanges between water and sediment. As already stated above, in most calculations these reactions are not considered. The waste loads term includes the external input from waste discharging sources.

The model was applied in three steps. In the first step, the contributions of inflows through the English Channel and the North Atlantic Ocean, and from atmospheric deposition, were estimated. For the inflowing waters this was done by applying a fictitious concentration to each of the Channel and North Atlantic inflows separately and making the corresponding runs with DELWAQ. Percentages of the locally present water masses, originating from the corresponding boundary, can be shown. This procedure was also applied to all riverine inflows and to the river Rhine separately. To derive the estimated concentrations in the planning area for a certain year, the resulting spatially defined concentrations can be multiplied by the ratio of the actual concentration in the considered flows to the fictitious concentration. Using a fictitious deposition rate of 1 $kg/km^2/year$, the same procedure was used for the inputs from atmospheric deposition.

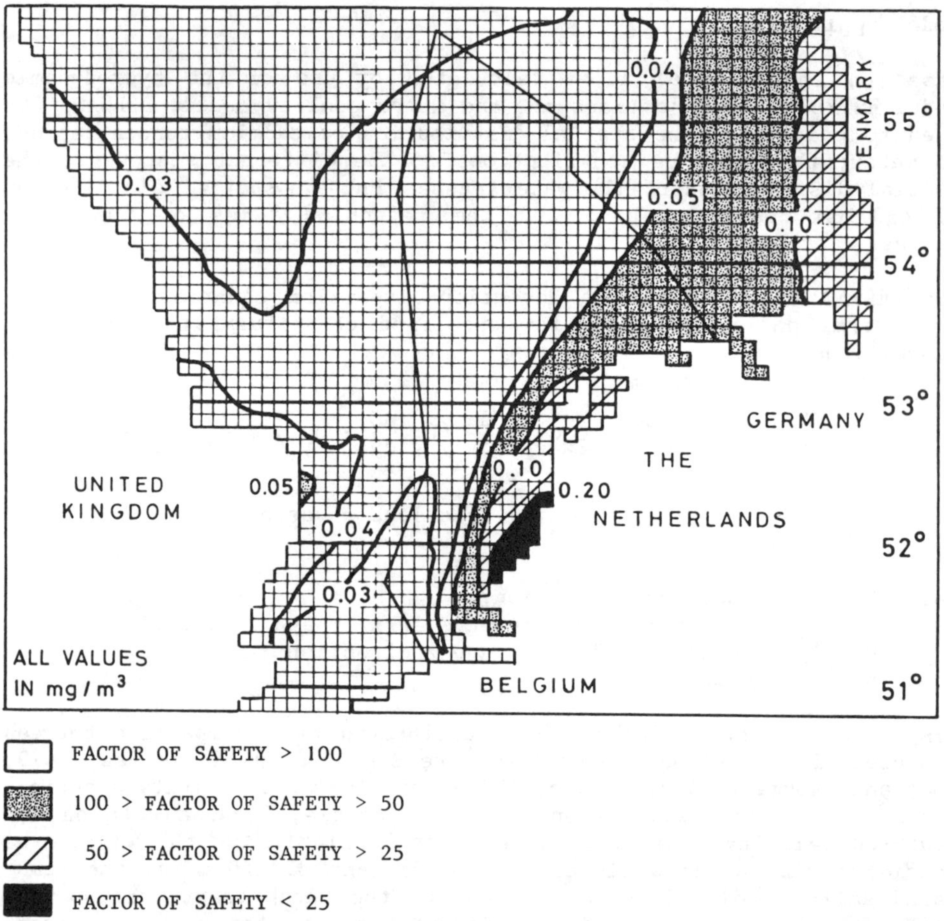

Figure 5-3. Estimated total cadmium concentrations and corresponding factors of safety in the southern North Sea under 1980 winter conditions, sedimentation not considered

In the second step, the contributions from riverine inputs and from direct discharges, i.e., coastal outfalls, dumping of dredging spoils, to the spatial distribution of cadmium concentrations were estimated. This was done by applying the estimated loads directly into the DELWAQ calculations at the respective input locations.

In the third step, concentrations in each grid cell were calculated by superposing, i.e., summing, the concentrations found in the separate calculations related to the different loads.

Results

Figure 5-3 shows the estimated distribution of total cadmium concentrations in the southern portion of the North Sea for 1980 winter con-

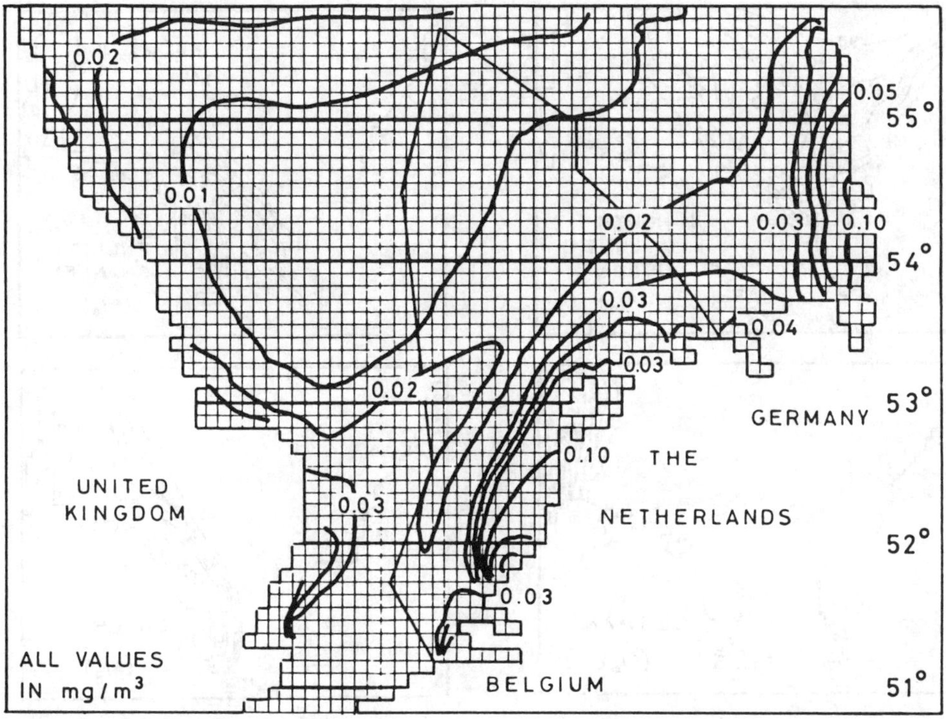

Figure 5-4. Estimated total cadmium concentrations in the Southern North Sea under 1980 winter conditions, sedimentation considered

ditions. Note that in no section of the region is the estimated concentration higher than the EC ambient quality limit of 2.5 mg/m^3 for dissolved cadmium in internal coastal and territorial waters. Figure 5-3 also shows the estimated factor of safety in each area, i.e., the lethal concentration, 5 mg/m^3, divided by the estimated concentration.

Cadmium concentrations calculated with the ESTFLO-DELWAQ model were compared with measured values as reported in Kooy and Stokman (1983), Duinker, et al. (1982), and Rijksinstituut voor Zuivering van Afvalwater (1983). These measured values were based on measurements in the Dutch coastal area, i.e., the area from 10 to 40 kilometers from the coast, where concentrations would be expected to be the highest. Estimated concentrations ranged from 0.04 mg/m^3 to 0.20 mg/m^3; measured concentrations ranged from 0.09 mg/m^3 to 0.15 mg/m^3.

As stated above, the concentrations estimated by application of the model, shown in Figure 5-3, ignore sedimentation processes. Figure 5-4 shows estimated concentrations in North Sea waters using a first order sedimentation velocity with K, the sedimentation rate, equal to 0.01/day. That is, every day 1% of the cadmium in the water phase disappears by sedimentation. Comparing Figure 5-4 with Figure 5-3 shows that including a reasonable estimate of sedimentation processes results in considerably lower estimated total cadmium concentrations in the southern North Sea.

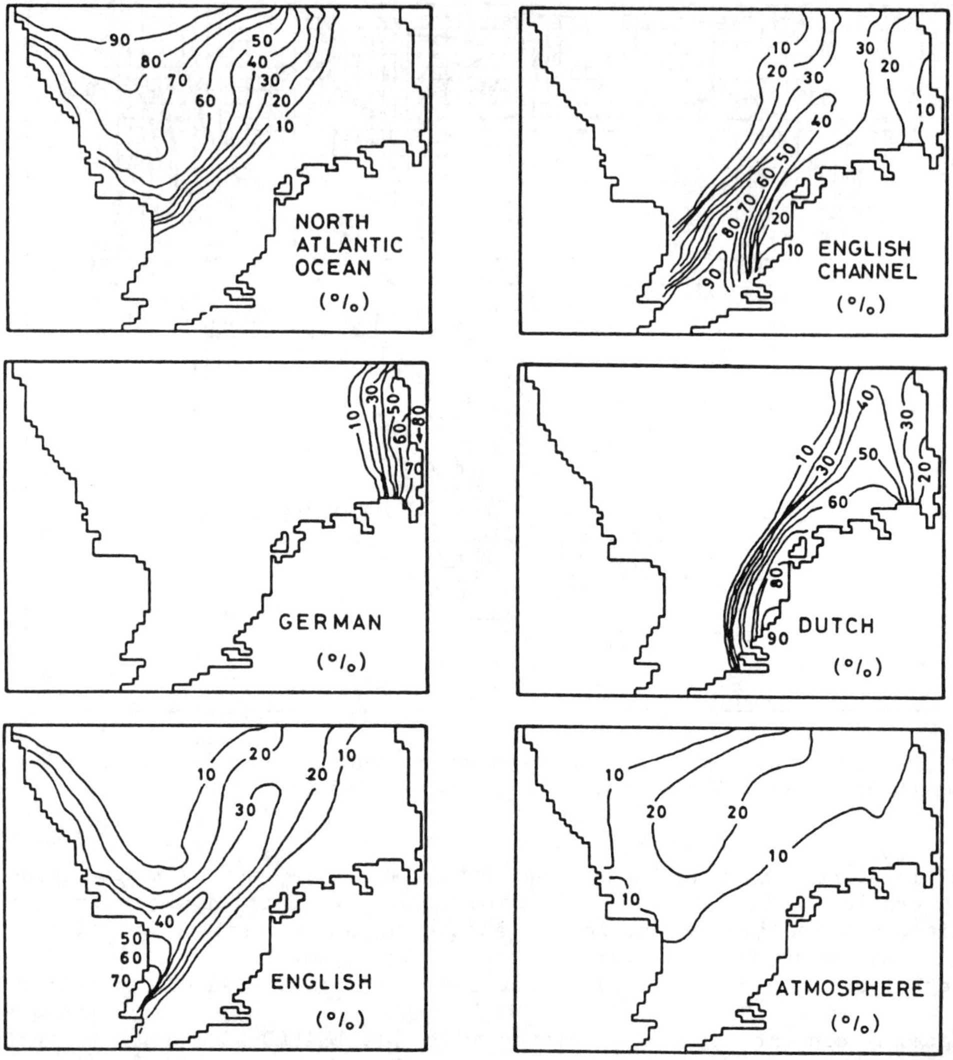

Figure 5-5. Estimated proportional contribution of different sources to cadmium concentrations in the southern North Sea for 1980 winter conditions, sedimentation not considered

Concentrations under average summer conditions will differ from the concentrations shown in Figure 5-3. In the ESTFLO-DELWAQ model, summer and winter conditions differ only with respect to the Channel inflow; in summer the inflow is assumed to be 100,000 m^3/s, in winter 200,000 m^3/s. As the influence of river inflows will be relatively higher in summer, estimated cadmium concentrations will be higher. However, lower wind velocities during the summer will increase sedimentation, resulting in lower concentrations in the water phase and higher concentrations in the sediments. It is not known what the net effect of these two factors is. The effects of lower river inflows in the summer will depend on the seasonal distribution of cadmium discharges into internal surface waters. This distribution, however, is not known.

In developing a strategy for managing ambient water quality, it is essential to know not only the spatial distribution of the absolute concentrations of a given material of concern, but also the sources of the loads which contributed to the concentration at any location in the region. Figure 5-5 shows the percentage contributions to the cadmium concentrations shown in Figure 5-3 of the following sources, as estimated by the ESTFLO- DELWAQ model combination: English Channel; North Atlantic Ocean; atmospheric deposition; and respectively German (FRG), Dutch, and English riverine and other inputs, such as coastal outfalls and dumping operations.

It should be emphasized that the identification of the source of input as the Dutch or German rivers does not mean that all of the load in the respective riverine inputs originates in the designated country. In the Dutch case, the load in the river Rhine as it enters the North Sea is comprised of loads from France, the Federal Republic of Germany, and The Netherlands. The load in the German rivers entering the North Sea is comprised of loads from Czechoslovakia, the German Democratic Republic, and the Federal Republic of Germany.

It is also critical from a management standpoint, to know to what extent ambient concentrations are attributable to man's activities, i.e., anthropogenic sources, rather than to natural, or non-human sources. The anthropogenic fraction is defined as:

$$A = (C - C_{nat})/C, \qquad (5\text{-}2)$$

where, A = anthropogenic fraction;
C = estimated cadmium concentration with loads from all sources; and
Cnat = estimated cadmium concentration using the estimated natural concentrations for the riverine inputs, the English Channel and Atlantic Ocean inflows, and atmospheric deposition, and excluding all other inputs.

The spatial distribution of simulated natural cadmium concentrations for winter conditions are shown in Figure 5-6. Figure 5-7 shows the spatial distribution of the estimated anthropogenic fraction of cadmium concentrations in the southern portion of the North Sea for 1980 winter conditions.

5.2 ESTIMATING EFFECTS ON CADMIUM CONCENTRATIONS IN SEDIMENTS

As stated in Chapter 2, cadmium is present in the North Sea both dissolved in water and adsorbed to suspended material and sediments. Cadmium adsorbed to suspended material settles in sedimentation areas and forms a cadmium containing sediment that may become a source of contamination for marine organisms, either through chemical processes resulting in dissolution of adsorbed cadmium or through direct consumption by benthic fauna. In estimating which part of the settled

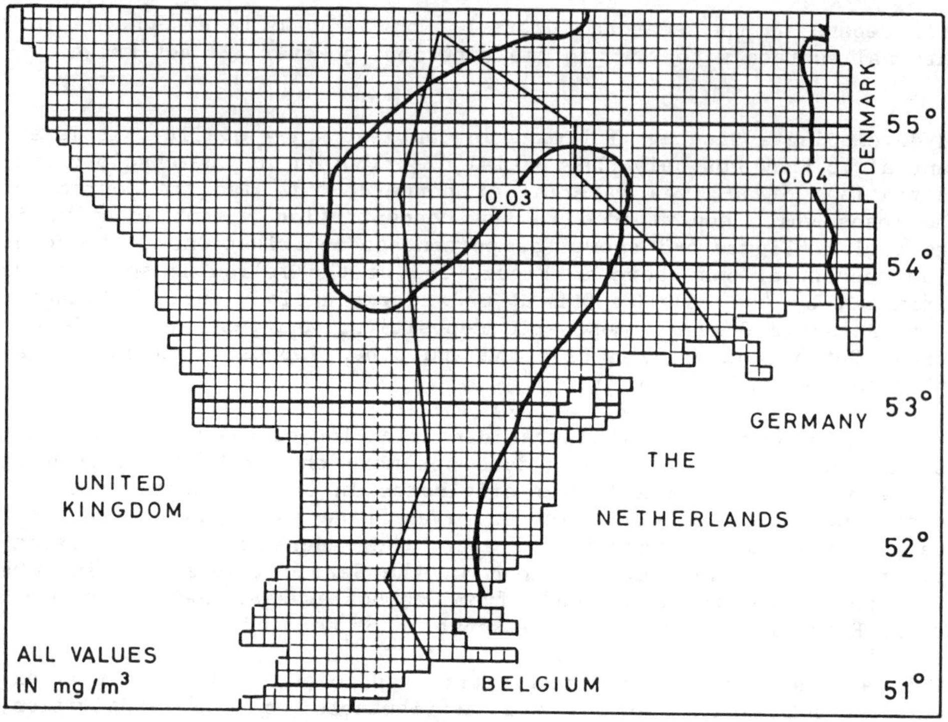

Figure 5-6. Estimated distribution of natural cadmium concentrations in the southern North Sea for winter conditions

cadmium may enter the food chain, man-induced, e.g., from dredging, and natural, e.g., by geomorphologic processes, perturbations of the bottom layer need to be taken into account.

However, knowledge of actual processes and of the actual situation is limited. In connection with developing the WQMP-North Sea, a preliminary analysis of sedimentation processes was made. Results of measurements of suspended material in Dutch coastal waters were analyzed in order to: define erosion and sedimentation areas; estimate erosion and sedimentation rates; and quantify related time scales. Measurements of suspended material were made in bi-weekly monitoring of water quality over a period from 1975 until 1983, at the 76 locations shown in Figure 5-8, as reported in Kooy and Stokman (1983).

Yearly mass balances were estimated for a 80 kilometer wide zone along the Dutch coast, based on quarterly residual flows calculated with ESTFLO. From this analysis it was concluded that erosion of sediments takes place in the southern part of the Dutch coastal waters, whereas sedimentation occurs in the Dutch delta region. Net sediment transport is directed to the Wadden Sea and to the German Bight area, north of the Wadden Islands.

To derive an estimate of the time scale involved in the accumulation of cadmium in bottom sediments in sedimentation areas, a first ques-

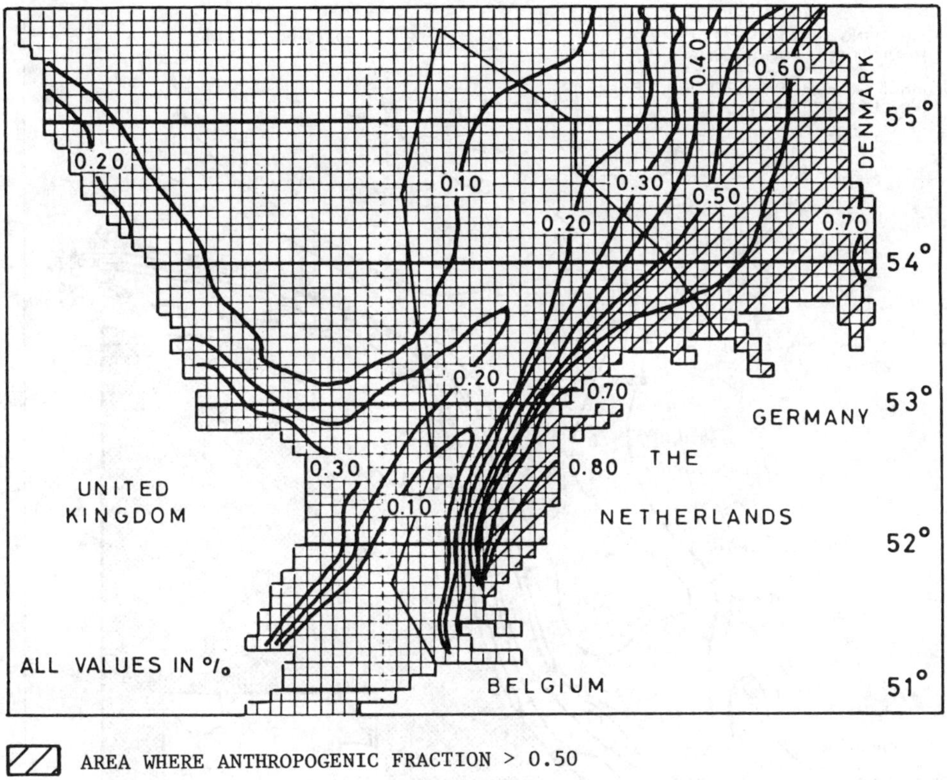

Figure 5-7. Estimated distribution of the anthropogenic fraction of the cadmium concentration in the southern North Sea under 1980 winter conditions, sedimentation not considered

tion to be answered is how the concentration of cadmium in the active surface layer of sediments develops over time as a function of the time dependent development of concentration of cadmium in suspended material from riverine and marine origins. The active layer is defined as the benthic layer which has an active interaction with the aquatic ecosystem. This interaction includes: physical activities, i.e., diffusion, perturbation by human activities, morphological processes; chemical processes, i.e., adsorption, desorption; and biological activities, e.g., by benthic organisms. Based on the morphological situation and the types of benthic organisms present, the active layer varies in thickness between 5 and 50 centimeters.

Assuming complete mixing between settled sediments and the receiving active bottom layer, the accumulation of cadmium in the sediments can be estimated (Anonymous, 1984) by:

$$\frac{dC_b}{dt} = \frac{S}{H} (C_{ss} - C_b) \qquad (5\text{-}3)$$

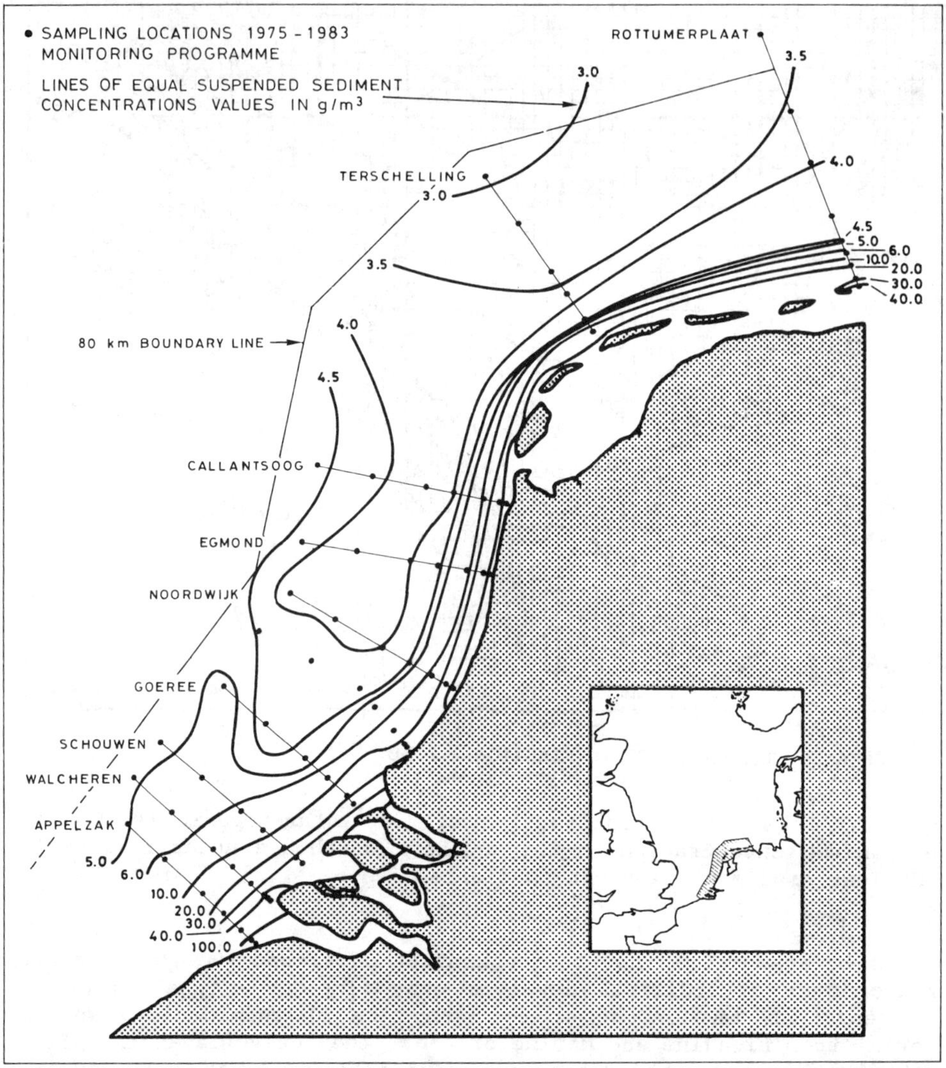

Figure 5-8. Suspended material sampling stations in front of the Dutch coast

where C_b = average concentration (over H) of pollutant in active surface layer, grams of pollutant per kg dry sediment;

C_{ss} = concentration of pollutant in suspended solids, grams of pollutant per kg dry sediment;

S = settling rate of suspended solids, mm/year; and

H = depth of the active layer, mm.

To estimate the trend in cadmium concentrations in sediments in Dutch coastal waters, it was estimated that the suspended material in front of the Dutch coast consisted of 90% marine material with a concentration of cadmium, Co, and 10% of Rhine sediments with a concentration

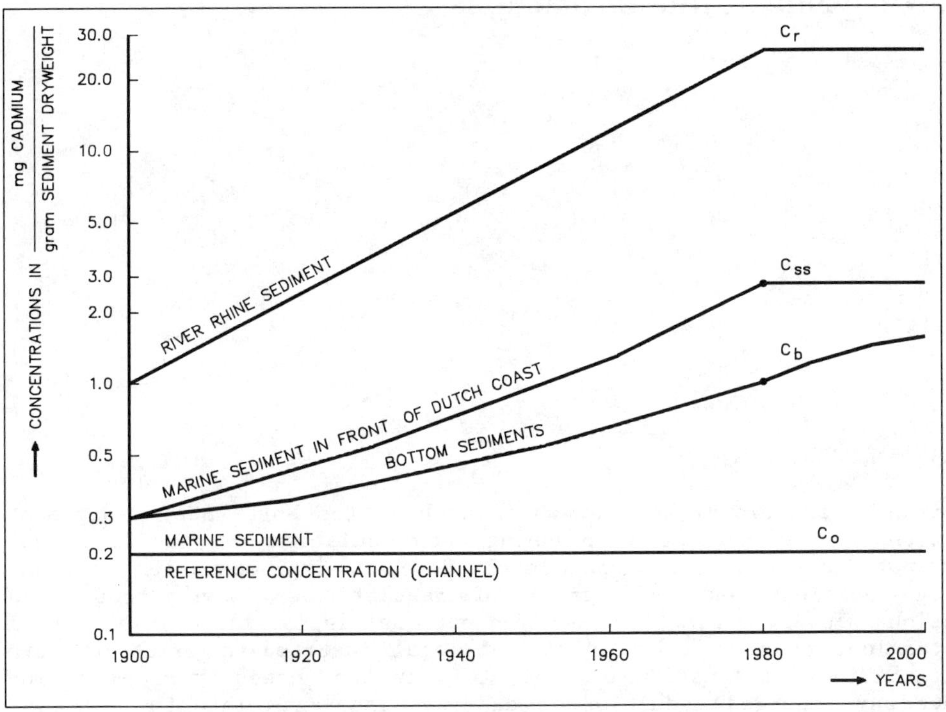

Figure 5-9. Estimated trend in cadmium concentrations in sediments

of cadmium of Cr. Based on measurements of Cr reported in Salomons and Groot (1978), the concentration Css can then be estimated. For a value of H/S = 50 years, which implies that --- after 50 years --- the total thickness of the settled sediments equals the depth of the active layer, the estimated concentrations over time in the bottom sediments, Cb, are shown in Figure 5-9. Figure 5-9 shows that, although the concentrations Co, Cr, and Css remain constant after 1980, Cb continues to increase.

In areas where the influence of the Rhine river is larger, such as in front of the coasts of South Holland and North Holland, the Rhine fraction in sediments is substantially larger than 10%. Therefore the cadmium concentration in the bottom sediments, Cb, will be higher than indicated in Figure 5-9. In 1982-1983 a survey was made to examine the quality of sediments along the Dutch coast (Salomons and Kerdijk, 1983). Cadmium concentrations were found in the bottom sediments (sediments < 63 micrometers) ranging from < 1.5 mg/kg to > 3.0 mg/kg. High concentrations were found in front of the coast of South Holland.

6 Analyzing institutional actions

6.1 INTRODUCTION

For an international body of water such as the North Sea, management actions --- in particular planning and regulation --- should be coordinated on an international level. However, implementation of planning decisions and enforcement of regulations always occur on a national level, subject to specific national legislation. In an actual situation, many administrative and legal instruments exist on both international and national levels to take the necessary steps to induce any required reduction of cadmium discharges into the environment. It was concluded in an analysis of these instruments (Rijkswaterstaat and Waterloopkundig Laboratorium, 1985a), that the major questions in this respect are not, which instruments are missing and what should be made available, but rather: how are the existing instruments being used; what triggers international and national actions; and how effective are the implemented actions?

In the first three sections of this chapter, a short analysis is given of the institutional actions on an international level, and on a national level with respect to the Netherlands and the United Kingdom, respectively. A short summary is given of the mechanisms available.

The last section contains a review of the reduction of cadmium discharges into Dutch coastal waters already achieved and of the additional reduction expected to be achieved as a result of actions currently underway. The latter actions, and some of the former actions, have been undertaken even though --- as indicated in the previous chapter --- for 1980 winter conditions, estimated concentrations of total cadmium in all grid cells were less than 10% of the EC standard of 2.5 mg/m^3 for dissolved cadmium in territorial seas and coastal waters. Given the fact, as described in the last section of this chapter, that cadmium discharges were reduced by about one-third between 1980 and 1985, the factors of safety in relation to the EC standard have increased substantially for many grid cells.

6.2 INTERNATIONAL MECHANISMS FOR WATER QUALITY MANAGEMENT

Triggered by specific disasters such as the Torrey Canyon tanker accident in 1967, by the establishment of "the environment" as a major political issue during the 1960s, and by a world-wide concern about environmental problems symbolized by the 1972 United Nations Stockholm Conference on the Human Environment, a number of international marine pollution control mechanisms were established in the seventies. Primarily these mechanisms were aimed at controlling vessel-source pollution. This source of pollution attracted attention because of the visible effects of tanker collisions and other accidents. Later, land-based sources were given much attention in an international framework, e.g., in the Paris Convention.

The general objective of marine pollution control mechanisms is to prevent marine pollution. Marine pollution is usually defined by international organizations as "the introduction by man, directly or indirectly, of substances or energy into the marine environment (including estuaries) resulting in such deleterious effects as harm to living resources, hazard to human health, hindrance to marine activities, impairment of quality for use of sea water and reduction of amenities" (Fotheringham and Birnie, 1979).

In most international mechanisms, e.g., conventions and directives relating to pollution, a distinction has been made between "black-list" and "grey-list" substances or groups of substances. Discharges of black-list substances should be eliminated, discharges of grey-list substances reduced. The former are considered to have substantially more adverse effects than the latter.

The scope of international regulations depends on how much agreement can be wrung from a collection of sovereign states. To establish a convention, all contracting parties which will be bound by the obligations of the convention after ratifying it, must agree on the contents. The same applies to the establishing of directives of the European Community. Because a sovereign state can be bound only by its own consent, the development of international pollution control measures is a lengthy process. A state which is the major disposer of a polluting substance can block any regulation which will involve it in what it considers to be unacceptable costs (Fotheringham and Birnie, 1979).

International conventions and directives have to be implemented in national laws and regulations. It is up to each party to decide what it needs to do to fulfill its enforcement responsibilities. Although EC directives are binding with respect to the results, member states may choose their own means.

Relevant international mechanisms in relation to cadmium inputs into the North Sea are the Oslo, Paris, and International Rhine conventions, and EC directives. They are described in the following paragraphs.

The Oslo Convention

This convention became operational in April 1974 and has been ratified by Belgium, Denmark, Finland, France, Germany (FRG), Iceland, Ireland, Netherlands, Norway, Portugal, Spain, Sweden, and the United Kingdom. The Oslo Convention refers to dumping from ships and aircraft, including incineration, on high seas and territorial seas. The Convention created a standing commission of representatives of the contracting parties to supervise and report on implementation.

According to the Convention the dumping of cadmium, one of the included black-list substances, is prohibited, except as a trace contaminant. On the basis of scientific investigations, a concentration of dissolved inorganic cadmium in seawater of 4.5 mg/m^3 has been defined, below which concentration adverse effects of cadmium are considered not to occur (Oslo Commission, 1984). Taking into consideration that some factor of safety is necessary to compensate for other possible toxic factors in waste discharges and that a minimum initial dilution of at least 1:1000 takes place within 5 minutes after discharge, a concentration of 1.0 mg dissolved inorganic cadmium per liter of waste has been defined as constituting a trace contaminant. All wastes containing dissolved inorganic cadmium in concentrations exceeding 1.0 mg/l are subject to a so-called Prior Consultation Procedure (PCP), according to which the secretariat of the Convention has to be informed of the origin, amount, and composition, plus other items, of the waste material before dumping. The PCP provides an opportunity to all contracting parties to object to the proposed dumping. Dumping titanium dioxide waste, dredging spoils, sewage sludges, and solid mineral wastes are not subject to the PCP procedure. Total cadmium concentrations in dumped material may never exceed the level of 4.5 mg/l.

The Paris Convention

This convention, entered into force in May 1978, has been signed by the EC and Luxembourg, and has been ratified by Belgium, Denmark, France, Germany (FRG), Iceland, Ireland, Netherlands, Norway, Portugal, Spain, Sweden and the United Kingdom. The Paris Commission, established by the Convention, is the central policy making agency in the execution of the Convention.

The Paris Convention refers to all pathways of pollutants in general, but more specifically is aimed at eliminating pollution of maritime areas --- high seas, territorial seas, estuaries, and internal coastal waters --- from land-based sources with substances on an included black list. Cadmium is one of the substances on this list. Programmes and measures for cadmium were adopted in 1985 and came into force on 1 January 1986. The Paris Commission adopted the limit values, water quality objectives, and all related provisions of the EC cadmium directive. (The directive is reproduced as Appendix B.) In the meeting of the Commission the delegations of Norway and Sweden stated that in their opinion more stringent quality objectives should be adopted. For the high seas, which waters were not addressed in the EC directive, the Paris Commission did not fix a water quality objective on the understanding that the quality objective for territorial waters will protect the high seas from pollution. A Dutch/German proposal for a cadmium standard of 1 mg/kg wet weight in mussels was rejected by the United Kingdom.

The International Rhine Convention

The International Rhine Commission elaborated, by order of the Second Conference of Rhine Ministers in Bonn (FRG) in 1973, the text of an international convention concerning the protection of the Rhine against chemical pollution. The Convention became operational on 1 February 1979, and has been ratified by Germany (FRG), France, Luxembourg, Netherlands, Switzerland, and the EC.

With respect to cadmium, the Rhine Convention expresses the willingness of the parties to end gradually cadmium discharges. Discharges of cadmium without a license, which specifies discharge standards, are prohibited. Limits for cadmium discharges, the same as incorporated in the EC cadmium directive, were adopted by the Rhine Commission in June 1983.

EC directives

The EC lays down environmental action programmes for periods of four years, e.g., 1973-1977, 1977-1981, 1981-1985, in which actions and measures with respect to environmental protection are specified. EC activities regarding the marine environment involve, in particular, promulgating directives and participating in important international commissions such as the Paris Commission and the International Rhine Commission.

The fact that measures aiming at protection of marine resources against pollution impose extra charges on national economies, which consequently might mean a deterioration of their competitive positions, is one of the grounds for environmental policies in the EC. According to the EC Treaty, EC directives can be established only with respect to a harmonization of national regulations directly influencing the common market. In the paragraphs of the preamble to the EC Directive on Aquatic Discharges (European Community, 1976), the following is stated:

> "whereas any disparity between the provisions on the discharge of certain dangerous substances already applicable or in preparation in the various member states may create unequal conditions of competition and thus directly affect the functioning of the common market .. whereas it is therefore necessary to approximate laws in this field".

Two EC directives are important in relation to cadmium. The first one is the so-called frame directive of 4 May 1976 (European Community, 1976), obliging the member states of the EC to take all adequate measures to <u>end</u> water pollution by black-list substances, i.e., those on List I. Although the original black list included in this directive mainly mentions groups of substances, instead of specific substances, cadmium and cadmium compounds were mentioned explicitly. The frame directive lays down that all discharges containing black-list substances require a license, in which discharge standards are defined. Such standards may not exceed certain limits, or may not result in concentrations in the receiving water body above certain levels. These values have to be defined in separate directives from the European Council, based on toxicity, persistency, and bioaccumulation. An interesting aspect is that, as long as no specific limits have been defined, black list substances are dealt with as so-called grey list substances (List II), of which the directive states that member countries are obliged to reduce water pollution.

Issuing the frame directive initiated the following procedure: (1) selection of substances to be regulated; (2) study of discharges and environmental characteristics; (3) design of a draft specific directive; and (4) decision making in the Council of Ministers. The first three steps are taken by the EC Commission, consisting of 14 members appointed by the governments, but then acting independently of national authorities. A large staff is under the jurisdiction of the EC Commission. The Commission decides on the progress of those steps. As soon as a draft directive has been submitted to the Council, the initiative is assumed by the European Council, composed --- for these directives --- of the ministers of all member states who are responsible for the environmental issues in their respective countries. The EC directives become operational after being accepted unanimously in the Council and officially published.

The first meeting of national experts to discuss the implementation of the frame directive for black-list substances was held in 1976. Cadmium was one of the substances which, with general consensus, was given priority for further action. After selection, the Commission contracted with consultants to carry out some eco-toxicologic, technical, and economic studies. Information about environmental effects, production, use, discharges, measures to reduce discharges, and costs of those measures was used in drafting the directive. In January 1982, the draft cadmium directive was discussed in the Council. However, the negotiations became stuck. As also was the case in the negotiations on the mercury directive, the bottleneck was the proposed regulation with respect to new industries. France, and other continental states, wanted to require the application of the best technical means available; the United Kingdom --- pursuing a policy based on water quality objectives --- would not accept this requirement (Dekker, 1983).

Finally, in September 1983 a compromise was reached. On September 26 the so-called cadmium directive was promulgated. Under Article 3 of the cadmium directive, all discharges of cadmium into internal surface waters, territorial waters, and coastal waters, require authorizations. These authorizations must contain discharge standards at least as stringent as those in Annex I to the directive, except when a member state can show in accordance with the monitoring procedure laid down in Annex IV, that the appropriate quality objective, presented in Annex II in the directive, is complied with in the area affected by the cadmium discharge. If necessary for the elimination of pollution or for prevention of distortion of competition, member states may grant authorizations for new plants only when those plants apply the best technical means available.

Standards were set for cadmium discharges in the following industrial sectors: zinc mining; lead and zinc refining; cadmium metal and non-ferrous metal industries; manufacture of cadmium compounds, of pigments, of stabilizers, and of primary and secondary batteries; and electroplating. Generally the limits are 0.5 mg/l in the discharge and 0.3 g/kg of cadmium handled, to be complied with as of 1 January 1986. Both limits should be applied to a given plant. No limits were established for the phosphate fertilizer industry. The directive states that no economically feasible technical means are available to reduce cadmium discharges in that industry.

The following quality objectives for receiving water bodies are specified in the directive: total cadmium concentration in inland surface waters, < 5 mg/m^3; dissolved cadmium concentration in estuaries, < 5 mg/m^3; and dissolved cadmium concentrations in territorial seas and coastal waters other than estuaries, < 2.5 mg/m^3.

According to Article 6 of the directive, the member states have to bring into force measures necessary to comply with the directive within two years following the notification, i.e., by September 1985.

6.3 NATIONAL MECHANISMS

The Netherlands

With respect to black-list substances such as cadmium, the approach of limiting discharges is adopted in the Netherlands. In the most recent Water Action Program 1985-1989 (Ministerie van Volkshuisvesting, Ruimtelijke Ordening en Milieubeheer and Ministerie van Verkeer en Waterstaat,1985), it is stated that, for black-list substances, surface water pollution, in principle, has to be terminated. The target is to reduce the discharges of these substances to as near nil as possible, by applying the best technical means available (BTM). If such BTM are not adequate in abating discharges, it has to be seen whether remaining discharges lead to unacceptable concentrations of cadmium in the aquatic environment. If this is the case, more far reaching measures are required, which may include a total ban on the discharges.

The above implies that BTM should be applied regardless of actual ambient environmental quality. In the Dutch approach, ambient water quality standards play a role only in assessing the situation which would result if BTM were applied. The Dutch ambient water quality standard for the total cadmium concentration in inland surface waters is 2.5 mg/m^3 in comparison to the EC standard of 5.0 mg/m^3. The Dutch government, however, considers neither the EC standard nor the Dutch standard to indicate the acceptable, safe, concentration. The Dutch standard is used to determine areas where measures to reduce cadmium concentrations have a high priority, i.e., areas where the cadmium concentration exceeds the standard of 2.5 mg/m^3. The Dutch government's view is that, in the long run, cadmium concentrations in surface waters should be near the so-called natural concentrations.

Implementation of international regulations occurs in the Netherlands through two acts which deal with black-list substances: the Surface Water Pollution Act (WVO) of 1970 and the Sea Pollution Act (WVZ) of 1977. Short descriptions of their fields of application and the instruments they provide follow.

The WVO is applicable to direct discharges to inland surface waters, territorial seas, and high seas through discharge pipes, and to discharges and dumping through other means to inland waters and territorial seas. The basic principle of this legislation is that discharging or dumping without authorization is prohibited. Permits will include conditions and instructions, which, in relation to cadmium as one of the black-list substances, imply the required application of BTM. Permits for discharges into state- managed waters, i.e., the major rivers and waterways, estuaries, and territorial seas, are granted by the Ministry of Transport and Public Works. Discharges into other waters are under the juridiction of the provincial executives, who are empowered to assign certain tasks and rights to the executives of other public bodies, e.g., the water boards. Through a ministerial decree in November 1974, a list of substances exists for which all discharges may be prohibited. No use has been made of this possibility.

Indirect dischargers, e.g., activities discharging into sewer systems, also must have permits. In general, such permits are granted by municipalities. In those permits, a discharge standard of 0.1 mg/l, not to be exceeded, is usually applied. This standard was specified in a directive of the Union of Water Boards. The Surface Water Pollution Act includes a list of types of activities which have to obtain permits from water boards or provincial agencies in order to discharge into sewer systems.

By virtue of the WVO, water boards responsible for ambient water quality apply a cadmium levy, which --- in 1980 --- was about 50 Dutch guilders per 100 g of cadmium discharged to waste water treatment plants. A cadmium levy can be imposed when extra treatment costs result from the presence of cadmium in the discharge to a treatment plant. A levy is meant to be a contribution to these costs, not an instrument to reduce discharges per se, although levies may have that effect.

With respect to discharges into surface waters from diffuse sources, such as agricultural drainage waters, other legislation has been proposed. The Soil Protection Bill and the Fertilizers Bill will be applicable to agricultural discharges. Standards will be laid down with respect to cadmium concentrations in fertilizers and sewage sludge applied on agricultural land. The Union of Water Boards has already established some standards regarding sewage sludge. Application of sludge on agricultural land is limited to 2 tons per hectare per year on cropland and to 1 ton per hectare per year on grassland. The cadmium content of the sludge must be less than 5 ppm. No standards have yet been established for cadmium in fertilizers.

The Sea Pollution Act (WVZ) concerns the dumping of wastes from ships and aircraft. The WVZ applies to Dutch ships and aircraft in all seas and to foreign ships and aircraft in Dutch territorial waters, including offshore installations in the Dutch portion of the Continental Shelf. Discharge or dumping of oil and radioactive materials was excluded, because other legislation already applies to these substances. According to the WVZ, it is absolutely forbidden to discharge into the sea any of the materials which are specified in an implementing regulation, such as cadmium, except where present as trace contaminants. The prohibition includes taking on board or delivering such materials with the intention to discharge them. The term "trace contaminant" has not been defined. An exemption permit is needed for discharges of harmful, polluting, or waste materials which are not specified in the black list, and for discharges of black-list substances present as trace contaminants. The Minister of Transport and Public Works, in concordance with the Minister of Housing, Physical Planning, and Environment, has the authority to grant exemptions from the prohibitions.

The United Kingdom

The water quality management approach of the United Kingdom seeks to strike a balance between the dual aims of protecting the environment and utilizing the natural capacity of the environment to accept and degrade many potential pollutants. In general the principle followed has been that primary responsibility for dealing with pollution problems should rest as far as is practicable with local authorities and water authorities, because it is they who are best able to determine what discharges are permissible in their areas and to identify any pollution that might occur. Thus, while national standards do exist in a few cases, in the main the central government's role is confined to laying down a statutory framework for regulation. Implementation is delegated to the local level, generally to the regional water authorities. In some cases advisory guidelines have been developed, but these do not have the force of law. There is in general no equivalent in United Kingdom legislation to the detailed and uniform discharge and/or ambient environmental quality standards in the legislation of many other countries. Authorities are expected to operate on the principle that standards should be reasonably practicable. Discharge standards can be established and are to be based on an assessment of local conditions and circumstances, the current state of technical knowledge, the potential harm from the discharges, and the financial implications (Department of the Environment, 1982).

The United Kingdom approach to dumping at sea is essentially a pragmatic one of resource management, based on the premise that the sea has an important role to play in waste management. Decisions with respect to licensing of discharges are designed to achieve the best reasonably practicable environmental option (Department of the Environment, 1982).

The main acts related to pollution control in the United Kingdom are: the Water Act of 1973; the Control of Pollution Act of 1974; and the Dumping at Sea Act of 1974. Discharges of cadmium into rivers, estuaries, and coastal waters are governed primarily by the Control of Pollution Act. This act prohibits the discharge of poisonous, noxious, or polluting matter, unless it is licensed by the relevant water authority. The ten water authorities were established under the Water Act, covering England and Wales. Dumping of wastes from ships and aircraft into the North Sea is regulated in the United Kingdom by the Dumping at Sea Act. The act creates a licensing system; to dump materials without a license is an offence. The Minister of Agriculture, Fisheries and Food is obliged in granting licenses to consider the need to protect the marine environment and its living resources. The minister must impose such conditions on the dumping as are necessary or expedient for such protection.

6.4 IMPLEMENTATION OF INTERNATIONAL REGULATIONS

In the Netherlands, the Paris Convention, the International Rhine Convention, and the EC frame directive have been implemented in the Surface Water Pollution Act. In particular, the International Rhine Convention and the EC directive have had major implications for this act. In amendments to the act, which went into effect 1 January 1982, the following were introduced: regulations concerning indirect discharges; the possibility of establishing limits; and the obligation to make an inventory of discharges.

The EC cadmium directive was also implemented in the Surface Water Pollution Act. The Dutch approach to abating pollution by discharges of black-list substances such as cadmium is directed to discharges. Therefore, in the implementation of the EC cadmium directive, the ministerial decree of September 1985, based on Article 1a.3 of the Surface Water Pollution Act which provided for establishing limits for discharge standards, adopted only the limits on discharges in Annex I of the EC directive. It was assumed that the cadmium concentration in waste waters never would be decisive. Therefore, only the limit values expressed in g/kg cadmium handled are incorporated in the Surface Water Pollution Act. Although the discharge standards at present applied in Dutch licenses are more stringent than the limits in the EC cadmium directive, the latter values were included in the ministerial decree.

The Sea Pollution Act constitutes the Dutch national implementation of the Oslo Convention. All provisions laid down in this Convention, such as the prohibition on dumping of certain substances, the exemption permits, and the black list, have been incorporated in the Sea Pollution Act.

To comply with the EC cadmium directive, the United Kingdom has relied upon administrative means. The Department of the Environment sent an advice note on the implementation of the directive to the water authorities. This note was drawn up in consultation with water authorities and industries. The advice note draws attention to the various provisions in the frame directive and in the cadmium directive and sets a time table for implementation. The note states that government policy is to consider ways of reducing exposure while taking into account the technical advantages of cadmium.

The provisions of the Oslo Convention and the goals set out in the Convention were embodied in United Kingdom law in the Dumping at Sea Act. It has to be noted, however, that the black list was not enacted as a part of that act.

6.5 DUTCH ACTIONS AND THEIR EFFECTS ON CADMIUM INPUTS

Because information on English cadmium discharges and on the way those discharges are being regulated is scarce, this section focuses on cadmium discharges into the Dutch portion of the North Sea.

Effects achieved through 1985

According to the Dutch Water Action Program 1985-1989 (Ministerie van Volkshuisvesting, Ruimtelijke Ordening en Milieubeheer and Ministerie van Verkeer en Waterstaat, 1985), cadmium discharges from industrial activities into Dutch internal waters decreased from 30 tons in 1975 to 19 tons in 1980 and to 18 tons in 1985. The major decrease is a result of reductions in discharges by fertilizer plants from 25 tons in 1975 to about 17 tons in 1980 and about 16 tons in 1985. Fertilizer plants discharge waste gypsum under licenses granted by Dutch authorities. Discharge standards in these licenses are based on the cadmium content of the raw material processed. Use of cleaner phosphatic rock was the measure adopted by the plants to reduce the cadmium content of the waste gypsum, in response to the discharge standards imposed.

Reductions of cadmium discharges also occurred in the electroplating industry, from 2.7 tons in 1975 to 0.4 ton in 1985. This was accomplished by improving wastewater treatment. Financial aid was available from the Dutch government. Individual electroplaters could claim a grant of 25% of the capital costs of a treatment plant and a grant of 40% of any consulting fees. After the reduction, the cadmium concentrations in discharged waste waters presumably comply with the standard of the Union of Water Boards, i.e., <0.1 mg/l.

An aggregate reduction of cadmium discharges of about 1 ton occurred in chemical and metallurgical industries by the end of 1985. Utilization of cleaner rock reduced inclusion of cadmium in pigments and stabilizers, and extraction of suspended particles from waste waters decreased cadmium discharges into surface waters. These measures have been taken in response to discharge standards imposed in the licenses issued to individual operations.

A considerable decrease in cadmium discharges into the North Sea via the Rhine River in the first years after 1980 can be ascribed to reduced cadmium discharges into the German part of the Rhine river basin. According to a study by the Delft Hydraulics (Waterloopkundig Laboratorium, 1986), the cadmium inflow into the Netherlands through the Rhine at Lobith was 110 tons in 1980, but only about 40 tons in 1983. The average flows were comparable, 2550 m^3/s in 1980 and 2650 m^3/s in 1983. The cadmium input into the North Sea via the Rhine River decreased thereby from about 64 tons in 1980 to about 32 tons in 1983. How the reductions by individual activities were induced, e.g., by imposing discharge standards, increase in factor prices, economic recession, combination of factors, is not known.

The cadmium input into the North Sea by the dumping of dredging spoils decreased, probably as a result of decreasing cadmium discharges into the Rhine. According to the Delft Hydraulics (Waterloopkundig Laboratorium, 1986), the dredging spoils from Rotterdam harbour dumped in 1981 contained about 36 tons of cadmium, the spoils dumped in 1983 about 22 tons. Estimated cadmium concentrations in the spoils were 23 mg/kg and 14 mg/kg, respectively. The total amounts of dredging spoils dumped in the two years were about the same.

The result of the above, plus other reductions, was an estimated reduction in cadmium discharges into the Dutch portion of the North Sea from about 136 tons in 1980 to about 89 tons in 1985, excluding atmospheric deposition.

Proposed efforts and anticipated effects

The Department of Public Works intends to prohibit waste gypsum discharges by the fertilizer industry at the beginning of 1988 (Luin, 1985). Recently the Dutch government tried to obtain financial compensation from the fertilizer industry for the extra costs resulting from the storage of polluted dredging spoils on land. However, the Dutch court decided (Anonymous, 1985) that the contribution of waste gypsum discharges to the pollution of dredging spoils was not sufficiently proven. The government is focussing its efforts with respect to the fertilizer industry on research. Grants from the national government are given for research on methods for utilization of waste gypsum and for the development of new technology for producing phosphate fertilizers.

With respect to cadmium discharges in dredging spoils, an additional reduction of about 2 tons is expected by 1990 as a result of decreasing cadmium discharges into the Rhine. From the beginning of 1987, only dredging spoils of Class I, i.e., the less polluted spoils from the areas near Hoek van Holland, and some portion of Class II spoils, may be dumped in the North Sea. Other dredging spoils will be stored on land. This shift in disposal method is expected to reduce the cadmium discharges in dredging spoils by an additional 9 tons by 1990, assuming there is no cadmium runoff from land disposal of the spoils.

Because storage on land is more expensive than dumping in the North Sea, the municipality of Rotterdam has initiated action to develop a long term solution to the contamination of dredging spoils with cadmium and other heavy metals. The intention is to charge additional storage costs to upstream dischargers, thereby inducing reductions of cadmium discharges. First steps have been taken to collect data on discharges through a survey in which the polluting activities, i.e., industrial operations and WWTP discharges, will be identified.

In addition to the storage of heavily polluted dredging spoils, the Dutch government has committed itself to two more actions. One is the elimination of the dumping of industrial waste, in 1980 containing about 1 ton of cadmium, by the end of this decade (Ministerie van Volkshuisvesting, Ruimtelijke Ordening en Milieubeheer and Ministerie van Verkeer en Waterstaat, 1985). The other is the installation of a biological waste water treatment plant near The Hague (Rijkswaterstaat, 1983). The latter action is estimated to reduce the cadmium load in waste waters and sludge discharges from 1.4 tons in 1980 to 0.3 ton in 1990. However, the main purpose for the installation of the treatment plant is not a reduction of cadmium discharges, but to reduce discharges of organic matter near the Dutch coast to comply with EC standards for swimming water.

The result of the above anticipated effects would be an estimated cadmium discharge into the Dutch portion of the North Sea in 1990, atmospheric deposition excluded, of about 76 tons, compared with about 89 tons in 1985, and about 136 tons in 1980.

7 Formulating management strategies

Producing outputs of products and services from a marine resources management region requires: (1) physical measures, i.e., technological means, for (a) producing the products and services, (b) handling, modifying, and disposing of wastes, (c) self-monitoring of discharges of private and public activities, and (d) monitoring of performance of activities and of ambient water quality by government agencies; (2) implementation incentives, which induce the activities to produce the desired mix of products and services from the region and to do so in a socially desired manner; and (3) an institutional arrangement which specifies (a) which governmental agencies and which private entities have responsibilities for which functions and subfunctions of marine resources management and (b) the modes of interaction between the public and private sectors involved in a marine resources management region. Thus, a marine resources management strategy for any given region has the following four components:

- the spatial locations and time patterns --- seasonal and over years --- of the products and services to be produced;
- a set of physical measures (technological means) for producing the products and services, for handling the wastes generated in production, and for monitoring performance in relation to the locations and time patterns of production and in relation to locations and time patterns for measuring ambient water quality;
- an implementation incentive system; and
- an institutional arrangement.

Although developing a management strategy for the North Sea in terms of multiple outputs is a logical objective, limitations of data and analytical resources precluded doing so in the ABC study, as discussed in Chapter 2. Instead, a more limited purview was adopted, namely, a focus on the use of the North Sea for waste disposal, and within that use of the North Sea, a focus on cadmium.

Strategies within such a limited context focus on the reduction of discharges, on zoning of activities affected by, or affecting, the concentrations of cadmium, or on both. For example, recreational or fisheries activities may be restricted to areas where cadmium concentrations do not exceed a certain level under average conditions. Such strategies have the same components as mentioned above for more broadly oriented strategies.

This chapter illustrates how one should logically go about generating information for decisions on marine resources management, within the context of reducing cadmium concentrations in the North Sea and/or limiting the utilization of the marine resources of this area.

In the first section an outline is given of the procedure to follow. Essential elements are estimating the costs and effects (benefits) of alternative strategies. (Ideally, the effects would be expressed in monetary terms.) When all benefits cannot be expressed in monetary terms, a cost-effectiveness approach can be adopted. In this section major attention is paid to an approach for assessing spatial differences in improved water quality in terms of beneficial effects.

The major components of a cost-effectiveness analysis are illustrated in the subsequent section, which deals with the estimation of the societal costs and cost-effectiveness of discharge reducing measures in the Netherlands.

Two additional sections contain discussions of: (1) implementation incentives, which comprise an integral part of formulating strategies; and (2) some practical considerations in the Netherlands related to the illustrated approach.

7.1 AN OPERATIONAL AND COMPREHENSIVE PROCEDURE

Two factors are of critical importance in formulating management strategies. One, implementating actions in the governmental sector typically must be done within budget constraints. Similarly, depending on economic conditions, the private sector may have limited capital and/or operating funds. Two, substantial uncertainties typically exist with respect to the results to be achieved in terms of improved ambient water quality and corresponding effects on marine organisms and human beings, by the installation and operation of various physical measures, given the complexity of an aquatic ecosystem such as that of the North Sea.

Given these factors, information which is of interest in the political arena includes: (1) what society must give up to achieve any specified target level of quality; (2) the incremental costs (and benefits) of obtaining improved water quality; (3) the distribution of costs, i.e., who pays; and (4) the distribution of physical and monetary benefits, i.e., who gains. The analysis must generate information on all of these, and the results must be presented in a manner understandable to the decision makers and to the other parties at interest, i.e., the various users of, and those interested in, the water body.

Formulating management strategies for marine resources regions has spatial and temporal dimensions. In a large body of water such as the North Sea, not all areas are of equal importance in relation to resources, uses, and management responsibilities. Nor are all areas alike in terms of their physical, chemical, and biological characteristics. Given the division of the area into units, such as the grid cells shown in Figure 5-2, the relative importance of subareas of the water body should be identified and explicitly taken into account in the analysis.

The temporal dimension has two facets. One relates to hydrologic and meteorologic variability; the other to economic variability. Both directly affect ambient water quality, the first by variations in natural inputs, the second by variations in discharges. As noted in Chapter 5, winter conditions differ from summer conditions in terms of residual flows into the North Sea. Riverine inflows also vary by season. Riverine inflows also vary from year to year, i.e., exhibit serial correlation with random elements. Tidal averaged inflows from the North Atlantic and from the English Channel are a function primarily of wind conditions, which can vary substantially from year to year.

Discharges vary as a function of economic conditions for an existing set of production facilities. Output in a year of depressed economic conditions can be as little as 50% of normal output; discharges are correspondingly decreased, although not necessarily in direct proportion. In addition, given that the output level and technology of activities change over time, discharges will also change over time as a function of those (and other) variables.

In the following section, the analysis procedure is discussed, with emphasis on the formulation of operational objectives and on the estimation of the costs and effectiveness of discharge reduction options. The management context is the single purpose, i.e., use of the North Sea for waste disposal, rather than the multiple output context.

Formulating the water quality objective

Analyses to develop management strategies are most often done for some assumed set of conditions, i.e., steady-state conditions with respect to both hydrologic/meteorologic variables and economic variables. Only one set of conditions may be analyzed, corresponding to the conditions in some base year. Or, several sets of conditions may be analyzed, reflecting conditions in the selected base year and in one or two or three future years, such as ten, twenty, thirty years beyond the base year.

Thus, for steady-state conditions, i.e., a specified spatial pattern of activities with associated levels of activities and discharges from those activities, the operational objective for a case such as that of disposal of cadmium in the North Sea, is:

> find the least cost set of measures and related implementation incentives to achieve a specified level of ambient water quality associated with a given factor of safety for a specified proportion of the subareas, i.e., grid cells.

The factor of safety implies a particular procedure for arriving at the target concentration, as described in Chapter 5. The target level may be derived in other ways. For example, the target level could simply be the estimated "natural concentration". Or it could be the concentration of chlorophyll alpha such that the color of the water is acceptable to the decision makers or to the users of the water. Or it could be a set of concentrations for different water quality indicators, e.g., dissolved oxygen, pH, temperature, alkalinity, deemed to be minimum levels necessary for some desired species of fish, such as the migratory shad in the Delaware and Connecticut rivers, the salmon in the Thames River, the mussels in the Eastern Scheldt. As noted above, the concentrations can differ from subarea to subarea.

However the concentrations are derived, the result is a three-dimensional operational objective, involving: concentration, % of spatial units, and costs, as shown in Figure 7-1. If the system were simulated over a multi-year period, for example using four seasons a year, a fourth dimension would be involved in defining the operational objective, namely, the proportion of time the target concentration is to be achieved in each subarea. Thus, the lower the concentration desired, the larger the percentage of subareas meeting the target concentration, and the larger the proportion of time the target is met in each subarea, the higher the costs. Given that the sociopolitical decision involves trade-offs among all four dimensions, expressing the operational objective in this form enables directly generating information on those trade-offs.

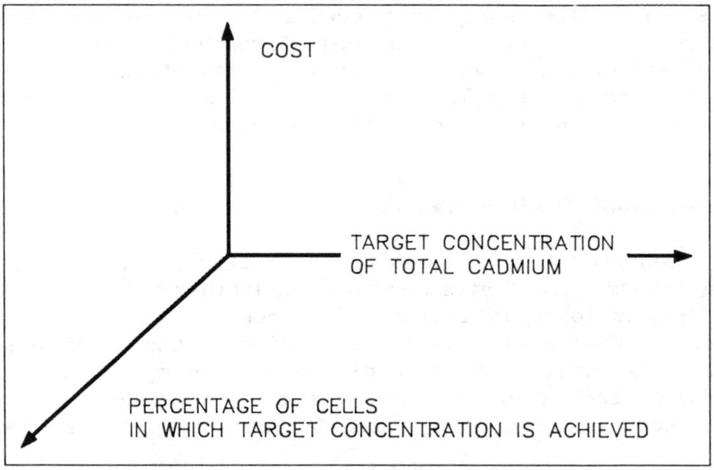

Figure 7-1. Dimensions of an operational objective for water quality management

The operational objective might also be expressed in these terms but with one or more types of constraints. For example, one type of constraint would be expressed: meet the target concentrations in 90% of the subareas, provided that the target concentrations were met in certain specified subareas. Or, the operational objective might be expressed: meet the target concentrations in all subareas subject to a constraint on total costs, i.e., capital costs plus operation and maintenance costs, or on capital costs alone. Where funds are limited, the analysis would seek to identify that set of subareas which would provide the largest benefits for the available funds.

Making target concentrations spatially variable in formulating the operational objective makes clear the importance and relevance of characterizing the relative importance or relative worth of different subareas of the region. Not all subareas of coastal waters, or of the open areas of the North Sea, are of equal value in terms of, for example, habitat for finfish or shellfish, or for recreation. Society may gain little for the resources expended by meeting or achieving the SAME concentration in all subareas.

The factor of safety is a simple, operational means to reflect BOTH the relative value of subareas and uncertainties in the analysis to develop management strategies. Uncertainties in projection of economic activities, uncertainty associated with water quality models, uncertainty associated with estimates of discharges, are examples. The factor of safety is used as follows, illustrated by fisheries as a desired output.

With respect to one fish species or several fish species, marine biologists would survey the subareas and classify them with respect to productivity, e.g., in terms of biomass of fish (kg) produced per cubic meter per year. Some areas with good density of subaqueous vegetation might well be highly productive. (Typically, a set of criteria is used for classification.) The result of this procedure would be that each subarea would be classified, for example, as high, fair, poor with respect to fish productivity. Logically, society would be most interested in preserving the most productive areas. Therefore, given the uncertainties in the knowledge of the effects of discharges on ambient water quality and of the effects of water quality on fish species, factors of safety would be assigned to each subarea in relation to its classification for fish production. Thus:

Classification (relative worth for fish production)	Factor of safety assigned
HIGH	100
FAIR	50
POOR	25

Then the factor of safety would be used to translate the lethal concentration, based on whatever data were available as described previously, into target concentrations for each of the subareas, in terms of cadmium concentration. The factor of safety is simply a means of expressing that society is more concerned about, and hence wants to be more sure of preserving, some areas than others.

The water quality indicators and shoreline characteristics relevant to determining the relative worth of different subareas depend on the output under consideration. Cadmium concentration has little relevance for recreational boating, or for hiking along the shore. Wind and wave patterns and shoreline topography are major determinants of such activities, respectively. For extractive outputs, such as oil and gas and sand, the magnitude and quality of the deposit in any given subarea are the determinants of relative worth.

Carrying out the analysis

Once the operational objective has been defined, the next step is to identify the sources of cadmium discharges and to estimate the quantities and time patterns of discharges, for the given levels of outputs and spatial pattern of activities. As stated above, this step is for each of the time horizons selected for analysis. This involves the projection of the types and levels of economic activities and households and household characteristics, where those activities will be located in the future, and the analysis of the factors affecting the discharges from those activities. Using these estimated discharges and some water quality model or set of models, the estimated concentrations resulting from these discharges, in both water column and sediments, in each of the subareas, are estimated.

When the specified concentrations in the specified proportion of subareas have been met, i.e., the operational objective has been achieved, then alternative operational objectives are specified --- both more and less stringent --- and the procedure is repeated. This enables estimating the incremental costs to achieve different levels of ambient water quality, and the effects of different constraints on costs and on ambient water quality.

Finding the least cost set of measures requires: (1) estimating the contribution of each significant source, point and non-point, to the concentration in each subarea where the target is not being met; (2) estimating the costs of reducing cadmium discharge from each significant source, i.e., a cost-discharge reduction table; and (3) selecting the least cost set of measures from among the options for the various sources. Because physical measures (technological means) to reduce discharges in the real world are discrete, a cost-discharge reduction table is relevant, rather than a cost-discharge reduction function. Both economies of scale and discontinuities preclude the existence of cost functions, i.e., continuous relationships, in most cases.

Because estimating costs is of critical importance in developing management strategies, some points relating to estimating costs merit mention.

- The costs are for one set of production levels and factor prices. Both production levels and factor prices can, and do, change over time, as economic conditions change.
- Costs are estimated using some base year for factor prices. Costs which have been estimated for other years must be inflated or deflated to the base year.
- Costs are generally expressed in terms of annual costs, in order to compare them with annual benefits, and include capital costs [K] and operation and maintenance costs [O&M]. Capital costs are converted to annual costs, for example, by use of a capital recovery factor [CRF], which reflects the assumed economic life of the investment and the assumed or selected discount (interest) rate, e.g., 15 years at 10%.
- All costs must include the costs of disposing of secondary, tertiary, etc., wastes generated in modifying the original waste streams.
- O&M costs for the discharging activity, public or private, include self-monitoring and reporting costs.

- O&M costs for the governmental water quality management agency include such costs as those for permitting, sampling, inspection of discharging activities and monitoring of ambient water quality.
- The proper basis for estimating the costs of reducing discharges by an activity is to compare the total production costs without the discharge constraint with total production costs for any specified degree of discharge reduction.

Using the data on contribution of a given activity to the cadmium concentration in any given grid cell, and the cost-discharge reduction table for the activity, a cost-effectiveness table for the activity can be derived. That is, the cost per unit reduction in cadmium concentration in cell X as a result of physical measure B by activity 07 can be calculated. By searching the cost-effective tables for all of the activities, the least cost set of measures to achieve the target in cell X can be determined.

The set of measures yielding the least cost to achieve the target in cell X will usually reduce concentrations in some other cells as well. The cost-effectiveness table for a given activity can then display the cost-effectivess of a given measure by that activity in relation to the target cell, and physical effectiveness in relation to other cells. It may be that the measure or set of measures which achieves the target level in the target cell will also result in achieving target levels in several other cells. Whether or not this is the case, there will be a new set of cells where the targets are exceeded. This will comprise the basis for the next "round" of the analysis.

The procedure is repeated until all of the specified management objectives have been analyzed. The resulting data show the "trade-offs" among level of ambient quality, geographic coverage, and costs. The flow of computational steps is schematically indicated in Figure 7-2.

Typically the analysis will start with the subarea which has the largest exceedance of the target, or the subarea which is considered by the decision-makers to be most important. After the target has been met in the first subarea considered, the analysis "turns" to whichever subarea now has the largest exceedance, or is considered to be the next most important. However, because of the complexity of coastal waters ecosystems, and because the multiple discharges are spatially separated and have different time patterns of discharge, it is possible that starting with a different subarea for the analysis would result in an overall, lower total cost to achieve the operational objective.

Computationally, the implication of the above is that the problem is to determine, from a population (set) of possible measures, that combination which SIMULTANEOUSLY achieves the targets at all locations specified, at least cost. At least three computational procedures are possible: (1) search driven by relative cost-effectiveness of the possible measures; (2) random sampling of the population of possible measures, leading to the selection of that combination in the sample which achieves the operational objective at least cost; and (3) some form of mathematical programming, i.e., linear or mixed integer. The random sampling approach is described in Hufschmidt (1962). Because of the limitations of mathematical programming methods, in terms of being

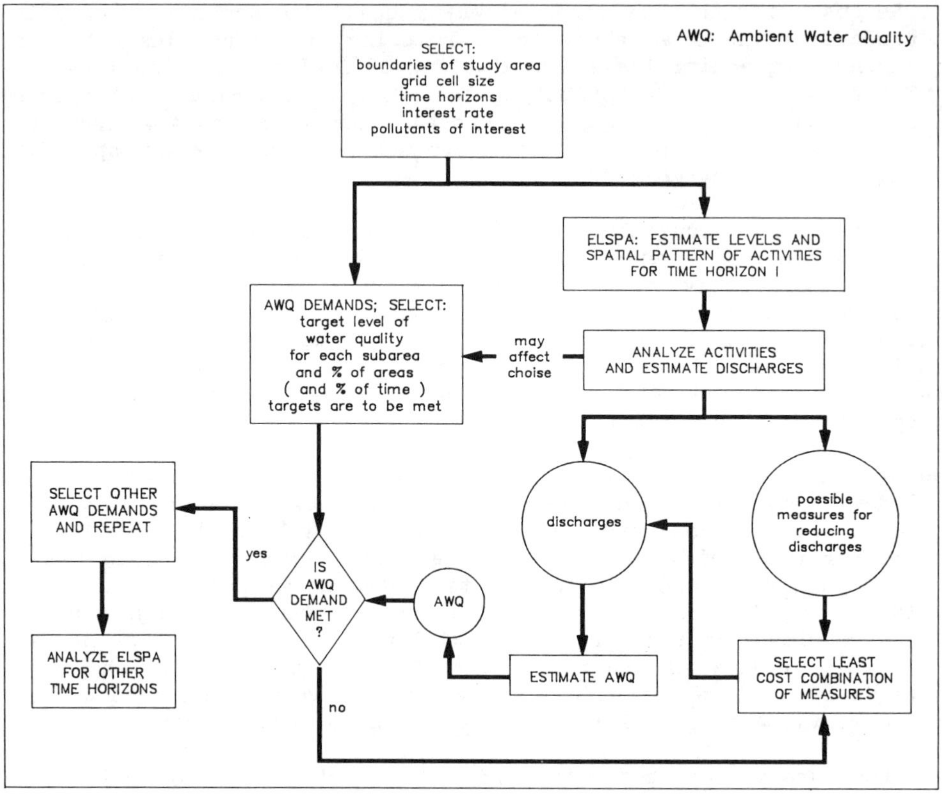

Figure 7-2. Computational steps in formulating strategies for managing cadmium for given levels and spatial pattern of activities

able to incorporate non-linearities, use of programming methods is likely to be possible in only a few cases.

In the final presentation of results, costs should be differentiated by activity groups, e.g., municipal wastewater treatment plants, dredging operations, chemical plants. These costs should be in terms of real resource costs to society, not costs to the activities. The latter will differ from real resource costs depending on various subsidies and tax provisions. Subsidies typically shift the costs from direct dischargers, public and private, to the public at large. Therefore, the final presentation should show both the total resource costs associated with the least cost system, and the distribution of costs for that system, given the existing set of subsidies. In some cases the least cost system will not be the desired system from a political point of view, because the distribution of costs is not acceptable.

However, it must be emphasized that cost-effectiveness of measures and the distribution of costs are not the only criteria used by decision makers in selecting a management strategy. In the last section examples of criteria for evaluating strategies are given. A few or many of these criteria, with varying weights, may be used in any given decision context.

7.2 ILLUSTRATION OF THE PROCEDURE

The procedure is illustrated in this section paying special attention to identifying critical areas and determining cost-effectiveness of possible measures. Subject of this illustration is the formulation of Dutch management strategies with respect to cadmium discharges in the Dutch portion of the North Sea.

Given the fact that there are multiple non-Dutch sources of cadmium discharges to the North Sea, an additional step in the analysis would be determining whether or not activities which can be undertaken by the Dutch would, by themselves, result in achieving the targets considered desirable by the Dutch for the Dutch portion of the North Sea. Suppose that the Dutch reductions do not result in reaching the target levels. However, the factor of safety for each of the grid cells can be calculated. For example, implementing all possible Dutch reductions in cadmium discharges might yield a factor of safety of 70 instead of the desired 100 in the critical subarea. The sociopolitical question then is, is the achievement of that level worth the cost. Suppose instead that all Dutch reductions yielded a factor of safety of only 10 in the critical subarea. This would indicate that little would be gained by Dutch expenditures, and that substantial reductions by non-Dutch sources would be necessary if the concentrations desired by the Dutch were to be achieved.

Critical areas

Critical areas should be identified with respect to productivity and/or uses, and with respect to existing or potential unacceptable concentrations of cadmium. As an example, in the Dutch portion of the North Sea three different areas were identified in terms of productivity for fish and other marine organisms. Figure 7-3 shows the three areas and the factors of safety applied for those areas.

The first area, the coastal zone with a depth between 0 and 20 meters, is generally considered to be relatively very important, because many organisms are in a very vulnerable stage in their life cycles when they are in this area. The depth of 20 meters corresponds approximately to a distance of 20 kilometers from the Dutch coast, which represents a width of two grid cells in the model of the North Sea. This area is considered to be highly productive, so that a factor of safety of 100 is assigned for the analysis. The corresponding target concentration for the selected lethal concentration of 5 mg/m^3 then is 0.05 mg/m^3. The second area is assumed to be a less important area, for which the factor of safety assigned is 50. A factor of safety of 25 is assigned to the third area. Each of the grid cells shown in Figure 5-2 COULD thus have a factor of safety assigned.

Figure 7-3 also shows two specific areas, C1 and C2. These areas, in front of the Eastern Scheldt and the Wadden Sea respectively, can be considered to represent critical areas with respect to a major function of the North Sea, namely preserving estuaries by seawater inflow.

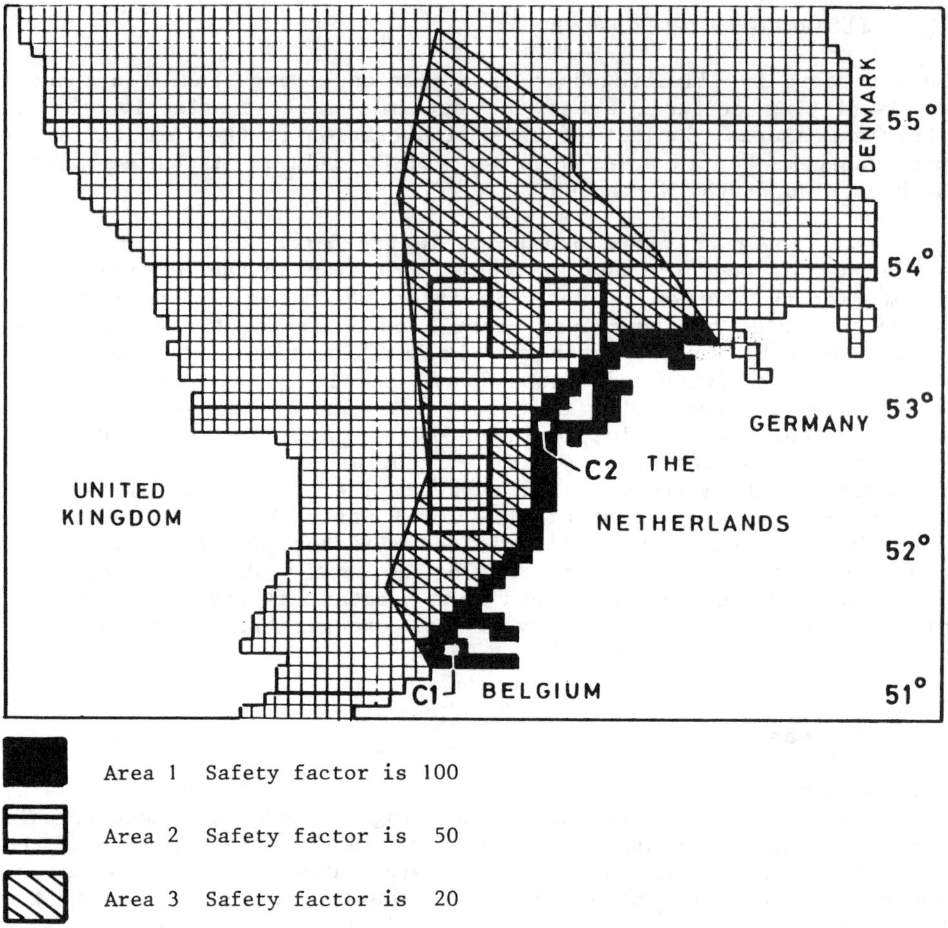

Figure 7-3. Example of delineation of relative importance of areas in the Dutch portion of the North Sea and corresponding factors of safety used in the analysis

Dutch measures for reducing cadmium discharges

In determining the cost-effectiveness of possible measures, the first step is developing a set of physical measures for reducing cadmium discharges by the various Dutch activities, and the associated costs.

In Chapter 4, the main Dutch sources of cadmium discharges in 1980 were indicated. To reduce cadmium discharges by Dutch activities, various measures can be undertaken, related to various steps in production processes and to use of products and services. Measures which reduce the total amount of cadmium discharged to the environment include: banning utilization of products containing cadmium; reducing the use of cadmium containing raw materials; and changing to other production processes which do not, or more efficiently, use cadmium. Measures which do not reduce the total amount of cadmium discharged to the environment but simply shift the discharge from one environmental medium to another include wastewater treatment and storage of dredging spoils on land, both assuming that containment is achieved. The following discussion illustrates how the discharge reduction-cost data relating to specific physical measures is developed.

Disposal of dredging spoil
The main Dutch dumping site for disposal of dredging spoil is Loswal Noord. At this dumping site, Site X in Figure 7-4, dredging spoils from the Rijnmond area are dumped. According to information presented in Chapter 4, the major part is classified as Class 1 (slightly contaminated), but some parts of Class 2 (moderately contaminated) and Class 3 (contaminated) dredging spoils are also dumped at this site. Of the dredging spoils being dumped, about one-fourth --- the Class 2 and Class 3 part --- is to be diverted to Site Y, where adjacent to the coast a special facility for disposal is being built. No leaching of toxic materials from the spoil is assumed to take place in this facility.

The difference in costs of dumping dredging spoil will be the difference between the cost of disposal of the entire amount at Site X, and the cost of disposal of 3/4 at Site X and 1/4 at Site Y. The mix of facilities and equipment will be different for the two systems. Because the discharge into the North Sea of undesired materials other than cadmium will also be reduced by this option, only a portion of the costs should be assessed "against" cadmium. One basis for allocating costs among the materials of concern would be on the basis of relative toxicity. However the allocation is done, it is important to emphasize that only a portion of the costs can be considered to be for the purpose of reducing cadmium discharges.

Also under consideration is moving the ocean dumping operation from site X to site Z, with the same proportions allocated to the two sites. The total amount of cadmium discharged would be the same as for the previous alternative, but the measure could have an extra positive effect on cadmium concentrations in the coastal zone, because of the current pattern along the coast. The extra costs would be substantial, because of the increased transport distance. As for site X, the cost to be attributed to cadmium would be some allocated portion of the net change in costs, other portions to be allocated to the reductions in discharges of other materials of concern.

Fertilizer plants
Fertilizer plants represent a major source of cadmium discharges. Conceptually, one alternative would be the addition of end-of-pipe facilities, analogous to waste water treatment plants. However, the volume of material involved would probably require facilities of such size as to make this option physically infeasible because of land constraints, and economically infeasible in any case.

Another alternative would be to cease fertilizer production and substitute imported fertilizer, or to use other types of fertilizer. The costs of both of these alternatives can be be estimated. Import substitution involves both direct and indirect costs. Consideration must be given to whether or not the plants can produce any other products with the existing capital investment, or with modified facilities. Whether or not other types of fertilizers can be used depends on the

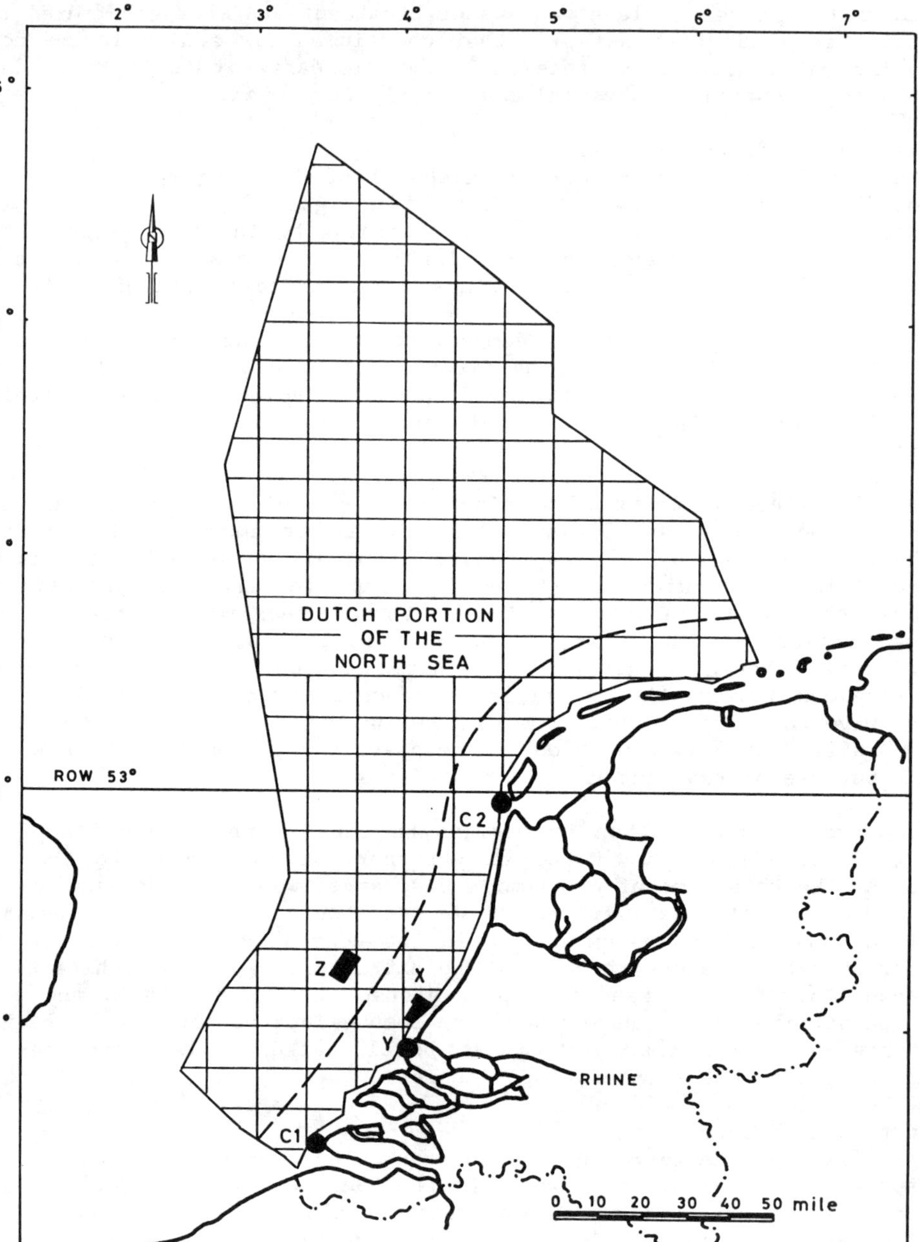

Figure 7-4. Existing and proposed sites for disposal of dredging spoils

nature and flexibility of the agricultural production functions involved.

A potential measure would be to use an alternative technology for producing fertilizer. One process has been developed through the laboratory stage. The new technology involves modifying the phosphate rock raw material, the direct input to the present fertilizer production process, to remove impurities, including cadmium. IF the new technology is technically and economically feasible, the costs to be attributed to reduction in cadmium discharge would be the difference in NET costs between the costs with the existing process, and the costs with the new process, including annualized capital costs for the new facilities. The new process involves a different combination of inputs, e.g., water, energy, labor, than the existing process, and has a lower product yield. It is also likely to generate a different set of liquid, gaseous, and solid residuals in production, so that there will be a difference in residuals modification and handling costs. There will be an increase in some types of residuals generated, e.g., energy residuals, because the new process is more energy intensive, i.e., requires more energy per unit of output. The costs for each production system must include the costs of handling and disposing of all primary, secondary, and tertiary residuals generated. There may be more potential for use, in byproduct production, of the waste slurry.

Agricultural operations

At least two alternatives could be considered. One would be more efficient application of the type of fertilizer currently in use. There is considerable evidence from a number of studies that farmers tend to use more fertilizer than is necessary to provide the required nutrient inputs for crops. This is done as an "insurance" measure. More efficient use of fertilizer possibly could be achieved by soil testing, monitoring content in relation to plant uptake, and more careful application. The net annual cost of this measure would be the cost of technical advice and monitoring minus the savings in fertilizer applied plus or minus the savings in costs of application. Unit costs will probably increase because of increased labor and equipment inputs to achieve the more efficient application. Total costs might decrease if savings in fertilizer costs are greater than the increase in application and other costs. The extent to which more efficient application of fertilizer would reduce discharges depends on the relation between application and runoff.

The other alternative would be a shift to a new type of fertilizer containing less cadmium, such as that which is presumed to be possible to produce by a new fertilizer production technology. Assume that the new fertilizer is available, but that it costs 30% more than the fertilizer currently used in agricultural operations. There is likely to be some response to change in price, i.e., elasticity of demand exists. Assume that the elasticity is such that the demand decreases by 15%. Therefore the new level of fertilizer application would be somewhat less than at present. This in turn would result in a decrease in cadmium discharge, even if the relationship between fertilizer application and runoff (discharge) remained the same as at present, which may or may not be the case.

The reduction in fertilizer use might result in a reduction in yield. If so, then the net cost of reducing cadmium discharges from agricultural operations would equal [reduction in yield X unit revenue] - net change in fertilizer application costs. The latter includes both the effect of the increase in price plus the change in application costs as a result of applying less fertilizer, remembering that the lower but more intensive application might result in an increase in unit application costs.

Analyzing costs in relation to some agricultural operations is complicated by the existence of crop support prices, and --- in some countries --- subsidized fertilizer and energy costs. Although estimating the resources costs associated with reducing cadmium discharges from agricultural operations is straightforward, estimating the incidence of those costs is often quite difficult.

Sewage treatment plant near the Hague
A decision has been made by the Dutch government to install secondary sewage treatment at the Hague. This decision was made for reasons other than reducing the discharge of cadmium. Even if cadmium in the North Sea were not an issue, the government would have gone ahead with the plant for other reasons. Therefore, unless there are particular, separable costs which would be incurred at the plant and in the subsequent disposal of sludge from the plant, which result from the presence of cadmium in the inflow to the plant, there are no costs associated with reducing cadmium discharges by the plant.

Chemical plant
At least two alternatives might be relevant for a chemical plant. One would be to add several steps of wastewater treatment, with no recovery of cadmium. The net cost of reducing cadmium discharge would be the total annual cost of the wastewater treatment plus disposal of wastewater treatment sludge, again assuming that the sludge was contained so that no cadmium would be discharged from sludge disposal.

Another alternative would be a shift in production process. Assume that this involves primarily addition and modification of unit processes in order to recover some portion of the used cadmium, thereby reducing discharge with the same level of product production (output). As with the fertilizer plant and new technology, many of the existing unit processes would still be utilized. The net cost then is the difference in the annual costs of production with the old technology and the annual costs of production with the new technology. The substitution of the new technology results in a different combination of factor inputs, e.g., energy, water, other raw materials, labor, so that the comparison must be made with respect to total production costs.

Summary of discharge reduction measures
The relevant data on the various physical measures would be summarized in a table such as Table 7-1. Table 7-1 also includes some measures with respect to minor cadmium discharges: discharges by the electroplating industry; discharges by the metallurgical industry; and dumping of industrial waste.

Tabulated for each activity would be one or more physical measures, plus, for each measure: capital (investment) cost, annualized capital cost, O&M costs, total annual costs, base discharge of cadmium (without the measure), discharge of cadmium with the measure, and reduction in discharge of cadmium as a result of the measure being in place and operating.

Cost effectiveness of physical measures

For any activity-physical measure combination in Table 7-1, the effect on cadmium concentrations by the reduction in cadmium discharge which would be achieved is estimated by use of the water quality model or models. These effects depend not only on the reduction in cadmium discharge, but on the discharge location as well.

With respect to discharge location, a relevant distinction to make is between: (a) discharges into the Rhine or the New Waterway, e.g., discharges by fertilizer plants and electroplating plants; and (b) discharges or dumping operations in front of the Dutch coast north of the Rhine outlet, e.g. the dumping of dredging spoils. Because of the current pattern along the coast, reductions in the latter type of discharges will only affect cadmium concentrations in areas north of the discharge location, i.e., in critical area C2 in Figure 7-4. Reductions in the cadmium load entering the North Sea through the Rhine may reduce the cadmium concentration in both critical areas C1 and C2, because Rhine water may be flowing southwards under certain weather and tide conditions.

Combining the change in concentration with the total annual cost yields the cost per unit of reduction in concentration, the "cost-effectiveness" of the measure. Compilation of such data for the various measures would be done as indicated in outline form in Table 7-2. This cost-effectiveness table can be used to find the combination of physical measures which would achieve sufficient reduction in concentration in the designated subarea to achieve the specified target concentration.

As noted previously, reduction in concentration in a specified subarea, i.e., grid cell, by a given measure is likely to reduce concentrations in other grid cells as well. Such information could be included in the cost-effectiveness table, as shown in Table 7-2.

To achieve the target levels in one or more cells in which the target levels are exceeded will usually require a set of physical measures. These combinations can be selected by searching the cost-effectiveness table or by random sampling, as indicated previously. Tabulation of the estimated results for the various combinations could be as shown in Table 7-3, in which the total annual cost for each combination is shown. The table also shows the grid cells of concern in each of which the resulting estimated concentration is less than or equal to the target concentration. In estimating the resulting concentration, it should be taken into account that not all reductions are physically independent, i.e., not all costs and effects are directly additive. Reductions in cadmium discharges into inland waters may affect the cadmium content of dredging spoils and, therefore, may affect the cost-effectiveness of contained disposal of dredging spoils.

Discharging activity	Physical measure	Resource costs, 10^6 1984 guilders				Estimated discharge in 1980, tons	Estimated discharge with measure, tons	Estimated reduction in discharge tons
		Investment cost (capital cost)	Annualized capital cost[a)]	Annual operation and maintenance cost	Total annual cost			
(1)	(2)	(3)	(4)	(5)	(6)=(4)+(5)	(7)	(8)	(9)=(7)-(8)
Chemical plant	Use raw material with lower Cd content					0.9	0.6	0.3
Fertilizer plants	Change production technology					17	3	14
Agricultural operations	Change to new fertilizer	0[b)]	0[b)]	0[b)]	0[b)]	4[c)]	1[c)]	3[c)]
Electroplating plants	Waste water treatment					1.5	0.4	1.1
Metallurgical plant	Waste water treatment					0.7	0.1	0.6
Municipal sewer system, The Hague	Install biological treatment	0[d)]	0[d)]	0[d)]	0[d)]	1.4	0.3	1.1
Disposal of dredging spoils	Change from open ocean disposal to contained disposal	e)	e)	e)	e)	36.5	20.2	16.3
Disposal of industrial wastes	Change from ocean disposal to land disposal	f)	f)	f)	f)	0.8	0	0.8

a) Annualized capital costs = capital cost x capital recovery factor (CRF). CRF = f(economic life, interest rate).
b) Assumes that use of new fertilizer will have no net impact on costs of agricultural operation.
c) Assumes that all agricultural operations will shift to new fertilizer and that there will be no change in amount of fertilizer applied.
d) No costs of the treatment plat are allocated to reduction on Cd discharges because decision to install was made for other reasons and because no separable costs atrributable to presence of cadmium appear to exist.
e) There could be several levels of this measure, each of which would reflect different amounts being shifted to special disposal area. Only some portion of costs shall be attributed to cadmium because measure will reduce discharges of other materials also, such as other heavy metals.
f) Only some portion of costs should be attributed to cadmium because measure will reduce discharges of other materials also.

Table 7-1. Activity - physical measure combinations to reduce cadmium discharges to North Sea from Dutch sources

Discharging activity	Physical measure	Total annual cost, 10^6 1984 guilders	Reduction in concentration in grid cell N17, mg/m^3	Cost per unit of reduction in concentration in grid cell N17, guilders per mg/m^3	Reduction in concentrations, mg/m^3, in grid cells:						
					N15	N16	N18	N19	N20	N21	N22
Chemical plant	Use raw material with lower Cd content										
Fertilizer plants	Change production technology										
Agricultural operations	Change to new fertilizer										
Electroplating plants	Waste water treatment										
Metallurgical plant	Waste water treatment										
Municipal sewer system, The Hague	Install biological treatment										
Disposal of dredging spoils	Change from open ocean disposal to contained disposal										
Disposal of industrial wastes	Change from ocean disposal to land disposal										

Table 7-2. Cost - effectiveness of physical measures to reduce cadmium discharges from Dutch sources

Combination	Resource Costs, 10^6 1984 guilders[a)]				Grid cells in Critical Area C1 in which concentrations ≤ target concentrations												
	Capital cost	Annualized capital cost b)	Annual operation and maintenance costs	Total annual costs	N15	N16	N17	N18	N19	N20	N21	N22	N23	N24	N25	N26	N27
(1)	(2)	(3)	(4)	(5)=(3)+(4)													
I				C	X	X	X	X	X	X	X		X	X			X
II				1.2C	X	X	X	X	X	X	X	X	X				X
III				0.9C	X	X	X		X	X			X	X			X
IV				1.5C	X	X	X	X	X	X	X	X	X	X	X		X

Notes: Combination I consists of: shift in location of disposal of dredging spoil, ——, ——, ——.

Combination II consists of: ——, ——, ——, ——.

a) Constraints on costs could be on capital cost of system, annual operation and maintenance costs, or on total annual costs.

b) Annualized capital cost = capital costs x capital recovery factor (CRF) = f (economic life, interest rate).

Table 7-3. Combinations of physical measures, their costs and effects

The foregoing examples are not meant to imply that these are the only ways of presenting the data. However, whatever the mode or format of presentation, the purpose is to show that the result of the analyses is the least cost combination of measures to achieve the operational objective.

The distribution of costs of a management strategy is often one of the major criteria used in sociopolitical decisions. Thus it is important for the analysts to estimate and tabulate not only total costs, but how those costs are distributed. The costs in tables 7-1, 7-2, and 7-3 represent resource costs to society. However, at least some of the activities are not likely to pay all of the resource costs, because of various subsidies. For example, in the case of the wastewater treatment plant for electroplating plants, the national government might provide a grant to cover 40% of the capital costs. For the new technology for fertilizer production, the national government might provide 50% of the development costs and 50% of the capital investment. The data on the distribution of costs for the least cost system can be displayed as shown in Table 7-4.

7.3 IMPLEMENTATION INCENTIVES

An integral part of formulating strategies is the analysis of implementation incentive systems. The purpose of an implementation incentive system is to induce desired behavior by resident and transient activities in a given marine resources region, such as the North Sea, or the Dutch portion of the North Sea.

An implementation incentive system consists of four components. These are:

(1) a set of rules or procedures an activity must follow, standards an activity must meet, production and/or waste modification equipment an activity must use, level of discharges which must not be exceeded, any or all of which are usually specified in an operating permit, license, franchise, lease;

(2) a basis and a procedure for measuring performance or adherence to the rules, procedures, standards specified in (1) for the activity, e.g., procedure for monitoring the quality of a liquid discharge over time;

(3) a schedule or procedure for inspecting the activity to determine compliance with the rules, procedures, standards, equipment specified, e.g., frequency of unannounced visits, submission of required reports in proper formats; and

(4) a set of sanctions, e.g., fines, imprisonment, revocation of operating permit, shutdown of production, for failure to comply with the rules, procedures, standards, et al., specified for the activity.

It cannot be assumed that, in the real world, the least cost set of physical measures identified will automatically be adopted and installed by the activities. Nor, based on data from various countries

Physical measures	Resource costs, 10^6 1984 guilders				Resource costs, 10^6 1984 guilders				Total annual costs, 10^6 guilders		Administrative costs, 10^3 guilders c)	
	Capital cost, K	Annualized capital cost b)	Operation and Maintenance costs, C	Total annual costs	Capital cost paid by activity	Capital cost paid by national government	O+M costs paid by activity	O+M costs paid by national government	Paid by activity	Paid by national government	Paid by activity	Paid by national governm.
(1)	(2)	(3)	(4)	(5)=(3)+(4)	(6)	(7)	(8)	(9)	(10)=(6)+(7)	(11)=(8)+(9)	(12)	(13)
Municipal sewage treatment plant					0.7K	0.3K	C	0	[0.7K][CRF] +C	[0.3K][CRF] +0		
Disposal of dredging spoils					0.3K d)	0.7K	0.5C d)	0.5C	[0.3K][CRF] +[0.5][C]	[0.7K][CRF] +[0.5][C]		
New technology for fertilizer production					0.6K	0.4K	0.8C	0.2C	[0.6K][CRF] +0.8C	[0.4K][CRF] +0.2C		

a) Distributions are illustrative only
b) Annualized capital costs = capital cost x capital recovery factor (CRF)   f(economic life, interest rate).
c) Administrative costs are generally so small that they would not affect total annual costs when rounded to closest 10^6 guilders. However, such costs must be budgeted; consequently it is useful to display them. Occasionally they will be large enough to affect the totals in columns (5), (10) and (11).
d) Funds obtained from harbour fees.

Table 7-4. Distribution of costs for physical measures in least cost combination

on performances by activities, can it be assumed that individual activities will automatically, continuously, adhere to their discharge or input or other specified conditions. Implementation incentive systems are essential.

Thus, for each activity-residual-physical measure combination such as shown in Table 7-1, the analyst must identify the set of incentives, such as subsidy for part of capital costs, permit, self-monitoring requirements, reporting requirements and inspections, which together will induce the activity to perform, and the governmental entities which will be responsible for imposing the incentives. For example, the national government might provide the subsidy, the provincial government issue the permit, and the provincial and/or local governments do the inspections.

The problem of implementation is particularly difficult in international contexts, as the discussion in the previous chapter shows. Differences in legal and governmental structures may mean that, for the same activity-residual-physical measure combination, different implementation incentive systems will be relevant in different countries.

The problem of implementation is typically ignored in analysis for formulating management strategies. Yet, without implementation the ambient water quality targets will not be achieved. In addition, there are some administrative costs associated with implementation, such as those relating to permitting and inspecting. Although these costs are usually not large in relation to investment and operating costs of facilities, the costs must be borne and hence need to be identified. For some activities in a marine resources region, such as discharging from vessels at sea, permitting and monitoring activities do not represent large costs to governmental entities. Monitoring of waste discharges from individual point sources, including inspecting, is likely to be substantially more expensive. In any case, the dischargers should bear the costs, as a part of the costs of doing business.

What the cost-reduction table in the previous section should then include, are the costs associated with implementation incentive systems, both the costs to the activity for monitoring and reporting, and the costs to the relevant agency for its tasks, e.g., permitting and inspecting.

7.4 PRACTICAL CONSIDERATIONS

Some remarks can be made with respect to (components of) the procedures set forth in the above sections.

- In the selection of measures to be taken, criteria other than the costs and the effects on cadmium concentrations in specific grids in the North Sea can be applied. Table 7-5 lists examples of criteria for evaluating strategies. Two comments with respect to criteria merit emphasis. One, it is difficult to quantify, rigorously, all of the criteria listed in Table 7-5. With respect to some, only qualitative scales such as low, medium, high, will be possible. For others, only descriptive commentary will be possible. Nevertheless, it is considered useful and essential to attempt to identify ex-

PHYSICAL, CHEMICAL, BIOLOGICAL EFFECTS OVER TIME
- a. Changes in ambient environmental quality in the various subregions and subsection of the region
- b. Effects of changes in AEQ on receptors, users of the region
- c. Extra-regional ecosytem effects
- d. Net energy used or produced

ECONOMIC EFFECTS AND THEIR DISTRIBUTIONS OVER TIME
- a. Direct benefits, e.g., values of products and services produced
- b. Direct costs of products and services produces
- c. Administrative costs
- d. Indirect benefits associated with products and services produced
- e. Indirect costs associated with proucts and services produced

ADMINISTRATIVE CONSIDERATIONS
- a. Simplicity
- b. Effects on staffs of existing agencies
- c. Retention of effectiveness under changing conditions
- d. Ease of modification under changing conditions, i.e., flexibility

TIMING CONSIDERATIONS
- a. Years before production of produces and services begins
- b. Years before adverse impacts on AEQ begin to be measured
- c. Time required to establish implementation incentive/institutional arrangement systems

POLITICAL CONSIDERATIONS
- a. Priority in relation to execution of strategies in other marine resources regions
- b. Degree to which strategy can be executed by a single agency rather than by multiple agencies
- c. Impacts on intergovernmental relations, i.e., relations between and among various governmental levels
- d. Acceptability to public
- e. Legal problems

ACCURACY OF ESTIMATES
- a. Physical, chemical, biological effects
- b. Benefits, direct and indirect
- c. Costs, direct and indirect

Table 7-5. Examples of criteria for evaluating strategies

plicitly, in any given context, the criteria of relevance for marine resources management decisions.

- Two, not only do the criteria affect the decision with respect to SELECTION of a strategy, they also affect the selection of analytical approaches and computational procedures used in the analysis. In order to APPLY the criteria to the proposed strategies, information on those criteria must be generated in the analysis, e.g., on the distribution of costs. Thus, the analytical approaches and the computational procedure and methods must enable the generation of the data needed to apply the criteria to proposed strategies.
- Some additional criteria may be relevant. Examples include: the reduction of cadmium concentrations in other (internal) waters or in other "compartments" (soil, atmosphere, food); and the reduction of concentrations of other polluting substances (such as organic pollution).

- It appears very difficult to obtain, in the Netherlands, accurate information on costs and effects of measures to reduce cadmium discharges. One reason is that some measures are still in development, such as a new production technique in the fertilizer industry. However, also in the case of measures already applied --- so costs must be known --- official reports do not include cost figures for individual private sector activities. The only information given consists of aggregated data on "environmental investments" in a certain industrial branch.
- The water quality model, as described in Chapter 5, is not sufficiently detailed for the kind of analysis outlined in this chapter, particularly with respect to estuarine areas. Grid cell size and structure of the model preclude analysis at that level of geographic detail. For example, as a result of the steady state conditions assumed in the model, and the existence of a wind and tide induced sea current from south to north, Rhine river water entering the North Sea always flows northwards. In reality, due to phase-differences between horizontal and vertical tidal movements, Rhine water may be flowing southwards during short periods of the tidal cycle, resulting effectively in stationary water masses with relatively little exchange with the open sea. In such areas sedimentation processes can cause accumulation of cadmium in bottom sediments.
- The fractional contributions of individual cadmium discharges in inland waters and the atmosphere can not be calculated, and --- even more important --- no distinction has been made between dissolved and adsorbed cadmium. This is an important question in relation to the assumed linear relation between the reduction of cadmium inputs into the North Sea from dumping dredging spoils and the reduction of cadmium concentrations in North Sea waters (Salomons, 1985).
- In a previous section a tool has been presented, by which spatial differentiation can be included in the analyses. However, although detailed information on biological characteristics is available, it is very hard to define the relative importance of subareas. Knowledge about the marine ecosystem still is limited, and particularly direct and indirect, short-term and long-term effects of substances, e.g., cadmium, on individual organisms and on the whole ecosystem are not very well known. Marine biologists therefore are very careful about making statements on the relative importance of subareas. However, by failing to do so, it remains for far less knowledgeable individuals to make such judgments, for the resources available to society are limited. Not all things can be done simultaneously, so that some allocation is necessary. Logically, primary allocation should be to those subareas which are more important to society.

8 Concluding comments

The concluding comments herein do not necessarily relate to substances other than cadmium and to aquatic systems other than the North Sea. Although they stem from analyses initiated in the Dutch context, at least some of them appear relevant to other situations. Seven categories of comments are discussed:

- international regulations, their triggers and effects;
- relating costs to benefits, internationally;
- implementation and control on national levels;
- specifying the water quality objective;
- analyzing the aquatic system;
- nature of the analysis; and
- remaining and further questions.

8.1 INTERNATIONAL REGULATIONS, THEIR TRIGGERS AND EFFECTS

International mechanisms for water quality management have at least the following two functions. One, they result in general statements on agreed-upon goals for reducing contamination of the environment. Two, they provide a "framework" for international deliberations on operational steps to be taken to reach such goals. The international mechanisms may contain concise and quantified standards, but situations to which they apply are not always defined explicitly. However, quality objectives and discharge standards established in international agreements do conceptually put a lid on concentrations and discharges, which may be important when, as a result of better economic conditions, production levels increase and pressures to increase discharges follow.

It is not known what information on the cadmium situation in the North Sea was used in drafting the international agreements, such as those discussed in Chapter 6. For example, no information could be found on the costs to specific activities in specific locations to reduce cadmium discharges; and only general information seems to have been used on the environmental effects of cadmium, not related to specific ecosystems in the North Sea or even to other aquatic ecosystems.

It is unclear on what quantitative information the decision of the EC was based to put cadmium high on the list of priority substances for reduction in discharges. Also, it is unclear on what information limits for ambient quality objectives and discharge standards in the EC cadmium directive, reproduced in Appendix B, were based. No definitions of "economically feasible" and "acceptable level", phrases used in the directive, are given. Was the acceptable level determined using the No Effect Level, or the Lethal Concentration divided by a factor of safety, or did it result from negotiations? Whether, and how, limits for ambient quality objectives and discharge standards are related, is not clear.

Answers to some of the questions raised above might well have been forthcoming if resources had been sufficient to enable digging into the EC files. The report of the contractor, minutes of deliberations of the Commission and of the Council, and any supporting information to the directive transmitted to member countries, might well contain information on the bases for various decisions reflected in the directive, and on the meanings of some of the terms used.

The Oslo Convention, the International Rhine Convention, and the EC cadmium directive seem to be the most explicit international agreements in existence, with respect to cadmium in the North Sea. However, it could not be shown that these international regulations contributed to the steps the Dutch government has taken, or intends to take. For example, with respect to the International Rhine Convention and the EC cadmium directive, measures to reduce cadmium discharges in the Netherlands were applied <u>before</u> these mechanisms went into force. Preventive influence seems unlikely because the Dutch government supported and supports a more stringent approach than the final compromises included in the EC directive. No discharge standards for fertilizer plants, one of the principal Dutch sources of cadmium discharges, are included in the EC cadmium directive; nor are there standards for discharges into the atmosphere.

The Oslo Convention excludes the dumping of dredging spoils from the Prior Consultation Procedure, permitting the Dutch to continue to dump a major part of the Rotterdam harbour sludge into the North Sea, even when cadmium concentrations in the sludge are above 4.5 g/m^3, the limit specified in the Oslo Convention. According to Table 4-2, cadmium concentration in dredging spoils from Rotterdam harbour in 1980 was 5.2 g/m^3.

One additional observation merits emphasis. Some of the reductions of cadmium concentrations in the North Sea result from measures taken to improve situations in inland surface waters, in particular in the downstream part of the Dutch waterway system, in relation to bottom sediments. In some other cases the reduction of cadmium discharges into the North Sea is a byproduct of the reduction of discharges of other substances. For example, the construction and operation of the waste water treatment plant near the Hague to comply with EC standards with respect to organic matter concentrations in swimming waters will also result in a reduction of cadmium discharges.

8.2 RELATING COSTS TO BENEFITS, INTERNATIONALLY

A typical problem in water quality management, in both intranational and international contexts, is that the activities and ultimately individuals who bear the costs do not necessarily, or rarely, receive direct benefits in proportion to the costs borne. The following are some examples. Not all individuals in households who pay more for costs of sewage treatment "use" the coastal waters for recreation. An industrial activity, such as a fertilizer plant, must incur expenses to reduce cadmium discharges into internal waters, which reductions will benefit downstream users but not the plant. Farmers who incur additional costs to reduce their discharges rarely receive benefits because their operations are upstream from the improved water quality. Typical direct benefits from improved water quality are reduced costs of intake water treatment, improved yields of various finfish and shellfish, and expanded opportunities for water-based recreation, e.g., enabling water-contact sports which previously were prohibited.

As pointed out in the previous chapter, the distribution of costs and the distribution of benefits represent important criteria for decisions in selecting strategies for water quality management. Two fundamental points merit emphasis. On the cost side, a basic principle is that all activities generate unwanted material and energy streams which require disposal. Disposal is generally to the environment, so that some amount of "environmental resource" for waste disposal is essential to any human activity. Steel, coal, electric energy cannot be produced without use of the environment. Therefore, the environmental media --- air, water, land as repositories --- represent necessary factor inputs into production and use, e.g., factories, farms, households. Therefore, the activity should pay for the use of the environment just as it pays for all other factor inputs, e.g., chemicals, energy, labor, capital. This is part of the underlying philosophy of the Polluter Pays Principle, espoused by the EC. Another part is that activities which result in damages to users of the environment should have to pay the costs of reducing or eliminating the damages.

Another important fact on the cost side is that, in a region of any size with multiple dischargers and different relevant subareas, the least cost system to achieve desired ambient water quality may well be one in which a large portion of the costs is borne by one or a few activities in a single country. Such a situation is often perceived as an inequitable distribution of costs, making agreement on action difficult.

On the benefit side, in addition to direct benefits there can be various types of indirect benefits. One type relates to a sometimes expressed societal objective, e.g., to maintain a given ecosystem, to "preserve the environment". Underlying that objective may be the philosophy that there are some intrinsic values to the environment, recognized by the bulk of the population of a country. Individuals and groups in the society may express themselves as being willing to pay something, i.e., to forego other goods and services, in order to provide funds/resources to pay the costs of improving ambient environmental quality. Such behavior reflects what economists term "option demand", the willingness of individuals in society to pay something to ensure that certain options relating to the environment will be available in the future, to succeeding generations, even if they themselves are not current, and will not be future, users of the environment.

The above discussion of costs and benefits is particularly relevant to a multinational context, such as that of the North Sea, where at least some of the benefits from improving ambient water quality are difficult to estimate in monetary terms, costs to achieve the imputed benefits may seem excessive, and the distribution of costs may seem inequitable. Various schemes for cost sharing might be proposed to reduce the "friction" among countries in arriving at a strategy for water quality management. For example, an "European Environmental Fund" might be established. Monetary sources for the Fund could be: (1) effluent charges on relevant polluting material being discharged, such as heavy metals; and (2) contributions from general revenues of the countries. The former reflects the principle that the polluter should pay; the latter reflects the principle that there are widespread benefits to society from improving environmental quality such that society in general should contribute to the costs. The international effluent charges would be analogous to a national government, such as the Federal Republic of Germany, imposing an effluent charge on an activity which already pays an effluent charge to a regional or state government.

Grants and loans could be made from the Fund to countries or individual activities to help pay for the costs of reducing discharges and improving ambient environmental quality, just as national governments make grants and provide below market rate loans to activities.

8.3 IMPLEMENTATION AND CONTROL ON NATIONAL LEVELS

Different nations have evolved different attitudes and approaches towards the presence of, and required measures against, toxic substances in the environment. Some countries, e.g., the United Kingdom, accept a level of ambient cadmium concentration which is higher than would occur without any discharges of cadmium in the environment, that is, higher than the natural concentration. Concentrations higher than the natural concentration are considered acceptable unless scientific proof can be given that such concentrations have adverse effects on the productivity of ecosystems and the health of human beings.

Other countries, such as the Netherlands, take the position that, considering the lack of knowledge of ecologic effects in the long term and considering the persistence of cadmium and its potential remobilization from sediments, it is not yet possible to define scientifically justified ambient water quality standards. For this reason they want to bring down the concentration of cadmium in the environment to what is considered to be a "natural" level. This view is in accord with the following decision of the Paris Commission (Paris Commission, 1984):

> "If the state of knowledge is insufficient, a strict limitation of emissions of pollutants must be imposed for safety reasons."

However, eliminating all cadmium inputs will not automatically result in natural concentrations in various water bodies. It may be expected that cadmium from human origin still may enter the water phase of the environment as a result of its prior accumulation in sediments and organisms.

The two approaches agree on the general goal: reduction of cadmium in the environment. They even agree on the means to reach such a goal, i.e., through reduction of cadmium discharges. However, the different philosophies about acceptable ambient concentrations seem to impede effective and efficient international action. Such action would require at least:

- agreement on the goals in concrete and quantified terms, e.g., on specific cadmium concentrations in various aquatic environments and on specific cadmium loads discharged to those environments;
- agreement on the data which represent an actual situation of discharges, ambient water quality, and effects on organisms;
- agreement on the methodologies used to estimate: (a) alternative future developments; (b) the effects of those developments on discharges of cadmium; (c) the effects of those discharges on ambient cadmium concentrations; (d) the effects of those concentrations on living marine resources and on other uses of the marine environment; and
- agreement on the costs and benefits, and on the distributions of the costs and benefits, of measures resulting from implementation of mutual agreements.

No in-depth analysis was made of national mechanisms, in the different countries involved, for enforcement of international agreements. The Dutch situation has been studied and, to some degree, mechanisms in the United Kingdom. It seems no more likely than in these countries, that detailed analysis in other countries would reveal that national mechanisms are insufficient to enforce any agreement laid down in international conventions. This implies that limitations to arrive at coordinated action to reduce the presence of cadmium in the environment are mainly of a political and not of an institutional nature.

8.4 SPECIFYING THE OBJECTIVE

A critical but essential step in setting up an analysis for water quality management, i.e., to develop a water quality management plan, is the specification of the water quality target or the water quality targets. Often the legislative intent is expressed in such terms as, "all streams and other surface water bodies should be fishable and swimmable by 19__"; "maintain the marine ecosystem"; "protect the environment". These are nonoperational, i.e., are not amenable to quantitative analysis. They must be translated into specific objectives/targets, expressed in quantitative terms. Failure to eliminate ambiguity results in difficulties for both the analysts and the decision makers.

As pointed out in the previous chapter, a relevant and an operational way of expressing a water quality target/objective is in terms of a combination of lethal concentration and factor of safety for each relevant spatial area, for steady-state conditions. Where time is also a dimension, the target then includes the percent of time the target concentration is to be achieved in each relevant spatial area.

In the study and the analyses for the WQMP-North Sea, inadequate attention was given to this specification problem, resulting in a lack of consistency. For example, as pointed out previously, the Dutch government has taken the position that what should be achieved in terms of ambient water quality is the "natural concentration". In the study to develop the WQMP-North Sea, effort was made to estimate the natural concentration of cadmium, both in the open ocean and in rivers. The results of the effort were total cadmium concentrations, i.e., dissolved plus adsorbed, of 0.03 mg/m^3 and 0.07 mg/m^3 for open oceans and rivers, respectively, as given in tables 5.2 and 5.3. of Rijkswaterstaat and Waterloopkundig Laboratorium (1985d).

Having made these estimates of the natural concentrations, and given the government's statements with respect to desired ambient water quality, it would seem that the logical procedure would have been to have adopted the above two concentrations as the target concentrations. The former would have been applied to the open ocean area of the Dutch portion of the North Sea, the latter to the near coastal area, e.g., grid cells reflecting depths to 20 meters. Instead, the stated procedure was to search the literature to find the "lowest lethal concentration" for any marine species, to which concentration a factor of safety of 100 was applied to calculate the target concentration. The result of this procedure was a concentration of 0.05 mg/m^3, used for all areas, i.e., all grid cells in the study area.

For estimated conditions in 1980, the target concentration used in the study of 0.05 mg/m^3 is exceeded in about one-quarter of the grid cells in the Dutch portion of the North Sea. No differentiation was made between grid cells in the open ocean and grid cells "near shore". In the period 1980-85, it was estimated that cadmium discharges to the Dutch portion of the North Sea were reduced by about one-third, as described in Chapter 6. As a result, the number of grid cells in which the target is exceeded would be expected to be reduced, although not likely in direct proportion to the reduction in discharges. What the results would be if the two different natural concentrations were used as targets is not known.

8.5 ANALYZING THE AQUATIC SYSTEM

The discussion in the previous section suggested the relevance of spatially differentiated targets. Spatial differentiation makes both ecologic and economic sense; the former because not all areas are of the same value ecologically, the latter because having spatially differentiated goals --- such that not all areas have to be at the same high level --- means lower overall costs to achieve the targets in all subareas. Moreover, natural concentrations may differ. Spatial differentiation, however, increases the complexity of analyzing the aquatic system, especially for a region as large as the North Sea.

Spatial differentiation is very likely to lead to scale problems. That is, near shore areas typically are more complex hydraulically --- and often ecologically --- than open ocean areas. Differences occur within much shorter distances. Therefore, analyzing the aquatic system of the area would require at least two scales, i.e., two different grid cell sizes, with the smaller being, perhaps 1/5-1/10 of the larger. The analysis then becomes an iterative one, in which analysis at the larger scale sets tentative boundary conditions for the smaller scale, and the analysis at the smaller scale --- which involves more detailed consideration of the physical, chemical, and biological processes --- provides the values for the various coefficients used at the larger scale, e.g., transfer coefficients. Iteration occurs until convergence to some specified "closeness" is achieved.

The above means: (1) the complexity of the analysis is substantially increased; (2) the cost of the analysis is increased, probably at an exponential rate; and (3) the accuracy of the results is increased, assuming that the model of the aquatic system at the small scale has been calibrated and validated. Whether or not the increased costs are justified depends on the value of the data generated.

8.6 NATURE OF THE ANALYSIS

The framework for analysis, and the characteristics of the analyses carried out in developing the WQMP-North Sea, were discussed in Chapter 3. The following are some major limitations of the analyses.

- Alternative mixes of products and services were not analyzed; only the one output, waste disposal, was considered. Only the levels and spatial pattern of activities existing in 1980 were analyzed, i.e., no analyses were made of projected levels of economic activities and of population.
- No specific water quality objective was defined in the beginning of the study. As the study progressed, two different routes were taken to determine an objective, resulting in the inconsistency described in the previous section.
- No specific measures to reduce discharges were formulated. As a consequence, no economic analysis was done of the costs of reducing concentrations in the North Sea. Similarly, no analysis was made of the incentives and institutional arrangements which would be necessary to achieve the specified ambient water quality.
- No analysis was made of the benefits of improved water quality. Even though it is difficult to express all benefits in monetary terms, at least they can be expressed in physical terms. This enables comparing, at a minimum, different mixes of physical outputs with costs.
- In the calculations with the water quality model, neither seasonal differences nor year-to-year differences were applied for river flows and discharges.
- Only total concentrations of substances analyzed were taken into account, rather than differentiating between those which were directly (immediately) biologically available and those which were not.
- Substances were considered to be conservative materials in the marine ecosystems.
- No sensitivity analyses were carried out with respect to the major assumptions, of which there were many.

The last point is well illustrated by the analyses to estimate the effects of discharges on ambient water quality, as described in Chapter 5. Given the assumptions made in developing and applying the models to estimate concentrations, and given the importance of the results in terms of decisions with respect to allocating society's resources, considerable analytical resources could justifiably be spent on analyzing the sensitivity of the results to the various assumptions. Such analysis would provide the decisions makers with some understanding of what the situation might be under, for example, the most conservative (adverse) set of assumptions. This would be useful in making decisions with respect to investment of public and private funds in measures to reduce cadmium --- and other --- discharges further. Sensitivity analysis would also provide an insight into which variables were more important, i.e., had the largest impacts on the results. Such information in turn would help in decisions with respect to research and data collection.

In Chapter 7 a more adequate, operational procedure to formulate a management strategy for water quality management (waste disposal use) was presented. The following are the main steps of this procedure. A more complete description of the steps in the analysis is given in Appendix A.

(1) Select the objective or alternative objectives, in terms of: factor of safety-target concentration-spatial location combinations; and the percentage of time each of combination is to be met. It should be emphasized that specifying different factors of safety for different locations is a first step in identifying and estimating benefits. These factors of safety are based on uses, and the relative magnitudes and importance of the uses, which are the basis for benefits.

(2) Identify sources of discharges for the time horizons selected, e.g., present, ten years in the future, twenty years in the future, and estimate the types, magnitudes, and time patterns of discharges from each of the sources, point sources, nonpoint sources, line sources.

(3) Estimate the time and spatial pattern of water quality resulting from those discharges.

(4) Determine whether or not the objective(s) has (have) been met.

(5) If the objective(s) has (have) not been met, determine the least cost set of measures to meet each objective.

(6) Present the results of the analysis in an understandable manner to show the trade-offs among: levels of ambient quality; proportion of area in which target is reached; percentage of time target is reached in each subarea; and total costs. Information on other criteria for selecting strategies must also be presented, such as the distribution of costs, the distribution of benefits, and the reductions in discharges of other substances.

Implementation incentive systems form an integral part of management strategies. Hence, formulating such systems must be included in formulating strategies. Components of implementation incentive systems are:

- a set of rules, procedures, standards applied to an activity, in terms of its performance;

- a basis and a procedure for measuring performance in relation to the rules, procedures, standards;
- a schedule and procedure for monitoring and inspecting activities to determine compliance with the specified rules, procedures, standards; and
- a set of sanctions which can be applied for noncompliance.

In determining the costs of measures to reduce discharges, the costs associated with implementation incentive systems should be included. These include, for example, costs of permitting and inspecting waste discharging activities, monitoring ambient water quality, imposing sanctions.

8.7 REMAINING AND FURTHER QUESTIONS

At least two important categories of questions still remain. One, are international regulations affecting behavior of countries with less stringent national regulations than the Netherlands, or do the international regulations represent the lowest (least stringent) possible common denominator of national standards? This is related to the preventive effect, i.e., inducing countries to take steps during the negotiations. Thus, is there any preventive effect on other countries? To examine this effect, standards applied in all EC member states or other contracting parties before, during, and after the establishing of the international agreement should be studied.

Two, the more adequate type of study described in Chapter 7 has been carried out in various contexts. The problem is how to convince politicians and agency bureaucrats that the type of study described is what is necessary in order to generate relevant information for decision-making. Both politicians and agency bureaucrats may have good reasons for not wanting more relevant information generated. In the last section of Chapter 7, some of the problems in trying to do the type of study described in the Netherlands were indicated. The following are considered to be the main problems in the Dutch context.

- In the selection of measures to be taken to reduce discharges, criteria other than the costs of measures and the effects of measures on concentrations in the North Sea, e.g., distribution of costs, can play a role. However, these criteria must be specified in the beginning, so that the analysts can develop methods to generate information on the criteria.
- It seems to be difficult to obtain, in the Netherlands, accurate costs on measures to reduce discharges by private activities. There apparently is no tradition of having private consulting firms experienced in various industrial sectors develop cost data for the government, as is true --- for example --- in the United States.
- The water quality model, described in Chapter 5, is not sufficiently detailed to enable accurate estimates of effects of various measures for reducing discharges on concentrations in the subareas of concern.
- Because knowledge about the North Sea ecosystem is limited, it is not easy to define the relative importance of subareas of the ecosystem. Yet, failing to do so implies that all subareas are of equal importance. This in turn will result in a much larger

expenditure of resources to achieve desired quality levels, than if a differentiation in importance is made. Only under a situation of no resource constraints, i.e., where there are no other ways to use the resources which would produce more benefits to society, might the uniform approach be reasonable.

References

Aalst, R.M. van, et al., 1983a, Vervuiling van de Noordzee vanuit de atmosfeer, TNO Report CL 82/152a, TNO, Organisatie voor Toegepast Natuurwetenschappelijk Onderzoek, The Hague

Aalst, R.M. van, et al., 1983b, Atmospheric deposition of lead and cadmium in the southern part of the North Sea 1, TNO Report R83/222, TNO, Organisatie voor Toegepast-Natuurwetenschappelijk Onderzoek, The Hague

Anonymous, 1984, Quality Status of the North Sea, Contributions of Member Countries to the International Conference on the Protection of the North Sea, 1984, Hamburg

Anonymous, 1985, Fabrieken Mogen van Rechter Gips Blijven Lozen op Waterweg, Volkskrant 85.11.23, Amsterdam

Baan, P.J., 1984, Baten milieubeleid water, Ministerie van Volkshuisvesting, Ruimtelijke Ordening en Milieubeheer, The Hague

Bower, B.T., Ehler, C.N., and Basta, D.J., 1982, Coastal and Ocean Resources Management: a Framework for Analysis, International Federation of Institutes for Advanced Study, Solna

Coordinatiecommissie Uitvoering Wet Verontreiniging Oppervlaktewateren, 1981, Aanbevelingen met betrekking tot de beperking van lozingen van niet- zuurstofbindende en toxische stoffen, die vrijkomen bij veel voorkomende processen in de metaalindustrie, The Hague

Dekker, R.H., 1983, Zwarte-lijststoffen, H2O (16), no. 12, pp. 274-279

Dekker, R.H., 1986, Rijksinstituut voor Zuivering van Afvalwater, Lelystad, Personal communication

Department of the Environment, 1982, Policies towards Waste Disposal - United Kingdom, Oslo Commission and Paris Commission, 13-14 June 1983, Berlin

Dijkzeul, A., 1981, De Waterkwaliteit van de Maas in Nederland in de Periode 1953-1980, Note 81.048, Rijksinstituut voor Zuivering van Afvalwater, Lelystad

Dijkzeul, A., 1982, De Waterkwaliteit van de Rijn in Nederland in de Periode 1970-1981, Note 82.061, Rijksinstituut voor Zuivering van Afvalwater, Lelystad

Directoraat-Generaal voor de Milieuhygiene, 1983, Verslag van het colloquium "Bestrijding waterverontreiniging door zware metalen" in Bunnik, 20 October 1983, Leidschendam

Duinker, J.C., et al., 1982, Dissolved copper, zinc and cadmium in the Southern Bight of the North Sea, Marine Pollution Bulletin 13(3), pp. 93-96

Eck, B. van, Turksma, E., and Sant, H. van 't, 1983, Voorstel referentiewaarden Nederlandse zoute wateren, Discussion Note, Ministerie van Volkshuisvesting, Ruimtelijke Ordening en Milieubeheer, The Hague

European Community, 1976, Directive 76/464/EC, Official Journal of the European Communities No. L129/23, Brussels

European Community, 1983, Directive 83/513/EC, Official Journal of the European Communities No. L291/1, Brussels

Fotheringham, P., and Birnie, P.W., 1979, Regulation of North Sea Marine Pollution, in: Mason, C.M., ed., The Effective Management of Resources, Frances Pinter Publishers Ltd., London

Gerritsen, H., 1983, Residual currents, a comparison of two methods for the computation of residual currents in the southern half of the North Sea, WL Report R1469-III, Waterloopkundig Laboratorium, Delft

Hufschmidt, M.M., 1962, Analysis by Simulation: Examination of Response Surface, in: Maass, A., Hufschmidt, M.M., Dorfman, R., Thomas, H.A. Jr., Marglin, S.A., and Fair, G.M, Design of Water-Resource Systems, Harvard University Press, Cambridge

Instituut voor Hygiene en Epidemiologie, 1983, Kwaliteitsoverzicht van een aantal Belgische oppervlaktewateren in 1982, Brussels

International Council for the Exploration of the Sea, 1979, Atlas of the Seas around the British Isles, Fisheries Research Technical Report no. 20, Copenhagen

International Water Tribunal, 1983, Casebook International Water Tribunal 3-8 October 1983, International Water Tribunal Foundation, Amsterdam

Kooy, L.A. van de, 1982, De Waterkwaliteit van de Westerschelde in de Periode 1964-1981, Note 82.063, Rijksinstituut voor Zuivering van Afvalwater, Lelystad

Kooy, L.A. van de, and Stokman, G.N.M., 1983, De Waterkwaliteit van de Noordzee 1975-1982, Note 83.084, Rijksinstituut voor Zuivering van Afvalwater, Lelystad

Koudstaal, R., 1987, Water quality management plan, North Sea: framework for analysis, A.A. Balkema, Rotterdam, 1987, IFIAS research series vol. 2, coastal waters No. 1

Luin, A.B. van, 1985, Rijksinstituut voor Zuivering van Afvalwater, Lelystad, Personal communication

Menzie, C.A., 1982, The Environmental Implications of Off-shore Oil and Gas Activities, Environmental Science and Technology 16, 8, pp. 454-472

Ministerie van Economische Zaken, 1983, Aardgas en aardolie in Nederland en op de Noordzee in 1982, The Hague

Ministerie van Verkeer en Waterstaat, 1982, Harmonization of North Sea Policies, Dutch Official Parliament Register 1981-1982, 17408, Numbers. 1 and 2, The Hague

Ministerie van Verkeer en Waterstaat, 1984, Harmonization of North Sea Policies, Dutch Official Parliament Register 1983-1984, 17408, Numbers. 7 and 8, The Hague

Ministerie van Volkshuisvesting, Ruimtelijke Ordening en Milieubeheer, 1984, Notitie aan de Tweede Kamer der Staten Generaal: Cadmium in het Milieu, The Hague

Ministerie van Volkshuisvesting, Ruimtelijke Ordening en Milieubeheer, 1986, Indicatief Meerjaren Programma Milieu 1986-1990, Environmental Action Program, The Hague

Ministerie van Volkshuisvesting, Ruimtelijke Ordening en Milieubeheer, and Ministerie van Verkeer en Waterstaat, 1985, Indicatief Meerjaren Programma Water 1985-1989, Water Action Program, The Hague

National Oceanic and Atmospheric Administration, 1984, National Coastal Pollutant Discharge Inventory: Gulf of Mexico component, Working Document on Procedures for Estimating Discharges, Office of Oceanography and Marine Assessment, Rockville, Maryland

Oslo Commission, 1984, Eighth annual report on the activities of the Oslo Commission, London

Oslo Commission and Paris Commission, 1984, The first decade, International co-operation in protecting our marine environment, London

Otto, L., 1983, Currents and water balance in the North Sea, in: Sundemann and Lenz, ed., North Sea Dynamics, Springer Verlag, Berlin

Paris Commission, 1984, Sixth annual report Paris Commission, London

Petrazzuolo, G., 1981, Preliminary Report on an Environmental Assessment of Drilling Fluids and Cuttings Released into the Outer Continental Shelf for the Gulf of Mexico, Environmental Protection Agency, Washington D.C.

Postma, L., 1984, A two dimensional water quality model application for Hong Kong coastal waters, Water Science and Technology, 16

Prandle, D., 1978, Residual Flows and Elevations in the Southern North Sea, Proceedings Royal Society London, Series A, 359, pp. 189-228

Rijksinstituut voor Zuivering van Afvalwater, 1983, De waterkwaliteit van de Rijn in 1982, note 83.080, Lelystad

Rijkswaterstaat, 1983, Waterkwaliteitsplannen voor Rijkswateren: Kustwateren (concept), The Hague

Rijkswaterstaat, 1985, Waterkwaliteitsplan Noordzee, The Hague

Rijkswaterstaat and Waterloopkundig Laboratorium, 1985a, Waterkwaliteitsplan Noordzee Achtergronddocument 1, Een Verkenning van Bestuurlijk- juridische Aspecten, The Hague

Rijkswaterstaat and Waterloopkundig Laboratorium, 1985b, Waterkwaliteitsplan Noordzee Achtergronddocument 2, Inventarisatie en Analyse, The Hague

Rijkswaterstaat and Waterloopkundig Laboratorium, 1985c, Waterkwaliteitsplan Noordzee Achtergronddocument 3, Activiteiten en Bronnen van Verontreiniging, The Hague

Rijkswaterstaat and Waterloopkundig Laboratorium, 1985d, Waterkwaliteitsplan Noordzee Achtergronddocument 4, Framework for Analysis, The Hague

Salomons, W., 1983, Zware Metalen in Aquatische Systemen: 1. Geochemische Inventarisatie; 2. Geochemische Processen; 5. Correlatie van de Gehalten in Water/sediment en in Organismen; 6. Bindingsvormen en Opname door Organismen, Rijkswaterstaat, The Hague

Salomons, W., 1985, Waterloopkundig Laboratorium, Delft, Personal communication

Salomons, W., and Groot, R.J. de, 1978, Pollution History of Trace Metals in Sediments, as Affected by the Rhine River, in: Krumbein, W.E., Environmental Biogeochemistry, Ann Arbor Science Publishers, Ann Arbor, Michigan

Salomons, W., and Kerdijk, H., 1983, Metalen in Boorkernen en Oppervlaktesedimenten van de Noordzee, WL Report M1676, Waterloopkundig Laboratorium, Delft

Salomons, W., Mook, W.G., and Eysink, W.D., 1981, Biogeochemical and Hydrodynamic Processes Affecting Heavy Metals in Rivers, Lakes and Estuaries, WL Publication No. 253, Waterloopkundig Laboratorium, Delft

Samenwerkende Rijn- en Maaswaterleidingbedrijven, 1983, Het Kwaliteitsprofiel van de Maas, 19-30 september 1982, Amsterdam

Stichting Natuur en Milieu, 1984, Cadmium, een Groeiende Bedreiging, Utrecht

Sundelin, B., 1983, Effects of Cadmium on Pontoreia Affinis (Crustacea: Amphipoda) in Laboratory Soft-bottom Microcosmos, Marine Biology, 74, pp. 203-212

Suylen, J.M., 1981, Quasi-synoptische Saliniteitsmetingen in de Zuidelijke Noordzee (Oostzijde), Report 0781 FA, Rijkswaterstaat, The Hague

Thomas, R., and Olsthoorn, A., 1984, Milieuverontreiniging door Pb en Cd, Instituut voor Milieuvraagstukken, Vrije Universiteit, Amsterdam

Train, R.E., 1979, Quality Criteria for Water, Environmental Protection Agency, Castle House Publications, Ltd., New York

Vliet, H.P.M. van, and Feenstra, J.F., 1982, Gebruik van Cadmium in Nederland, een Stofbalans voor 1980, Instituut voor Milieuvraagstukken, Vrije Universiteit, Amsterdam

Waal-Malefijt, A. de, 1983, Cadmium in Huishoudelijk Afvalwater en in af te Voeren Neerslag, Installatietechniek, 1983, 1, pp. 29-37

Waterloopkundig Laboratorium, 1984, Herkomst en Transport van Contaminanten op de Noordzee, Relatie tot de Dumping van Baggerspecie, WL Report S598/S599, Delft

Waterloopkundig Laboratorium, 1986, Zware Metalen Balans Rijn, fase 1: Inventariserende Studie, WL Report R2266-T138, Delft

Wit, J.A.W. de, Schotel, F.M., and Bekkers, L.E.J., 1982, De Waterkwaliteit van de Waddenzee 1971-1981, Rijksinstituut voor Zuivering van Afvalwater, Lelystad

Appendix A: ABC framework for analysis

Analysis is one of the basic functions of marine resources management and is an essential task in planning, i.e., the process of selecting a management strategy for any given marine resources management region. Because marine resources management is a continuous activity, the decision regarding the mix of products and services to produce on and from the region must be made from time to time (this is referred to as the planning decision). Hence, analysis for management must be virtually continuous, with specific outputs of information required at specific points in time. Analysis is that activity which produces information for the planning decision. Thus, analysis is a part of planning. The actual planning decision is normally part of the political process, undertaken after relevant information is generated in the analysis.

Analysis for planning should be clearly distinguished from research and data collection the objective of which is to increase basic knowledge, e.g., of physical, chemical and biological processes. Such research and data collection are essential for further developing the capacity to predict consequences of management actions, and therefore should be tuned to this prediction need. The delineation of both short-run and long-run data collection and research efforts should be an integral part of the planning process. The outputs should be programmed to become available to the planning process at specified points in time in the future.

Figure A-1 shows the seven segments comprising analysis for marine resources management. These segments are: (1) setting up the analysis; (2) estimating demands for products and services which could be produced in the region; (3) analyzing the potential activities and the so-called "combined" sources of discharges (rivers, groundwater, atmosphere); (4) analyzing natural systems including cross-boundary flows; (5) formulating and analyzing management strategies; (6) evaluating strategies; and (7) presenting results. Discussion of each of these follows.

Segment 1: Setting up the analysis (SUTA)

---

*) Excerpted from Koudstaal (1987)

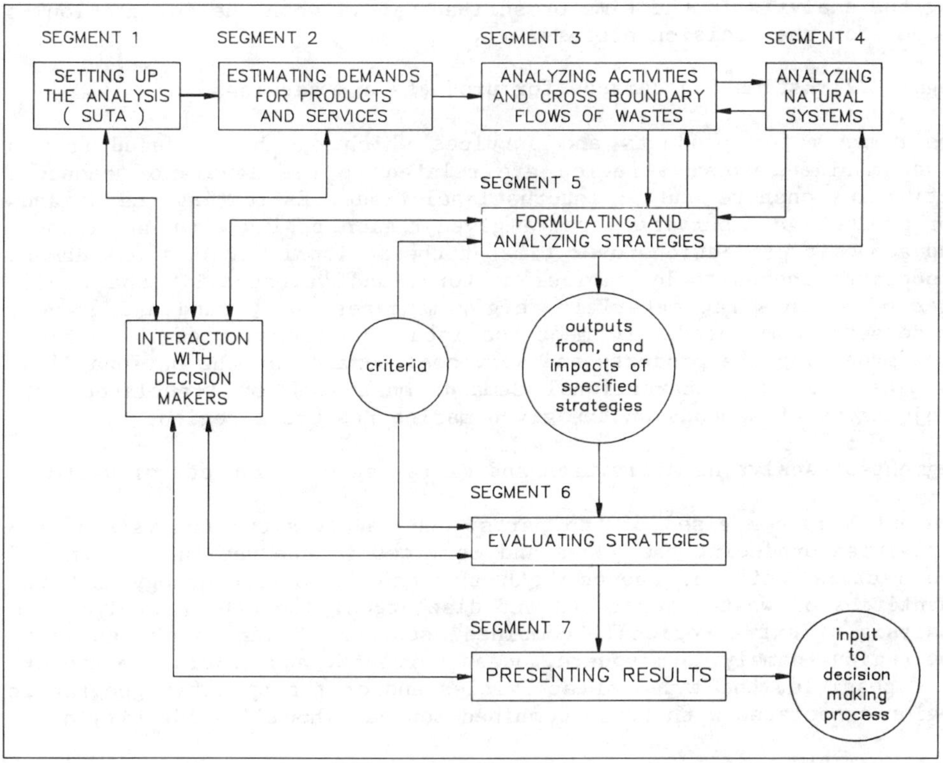

Figure A-1. Segments of analysis for marine resources management

SUTA refers to the sequence of steps - and related criteria, considerations, and assumptions - which must be performed to "set up" an analysis for marine resources management. Included are, for example, the delineation or preliminary selection of:

- the boundaries of the region and subregions to be analyzed;
- the time horizons or the time period to be analyzed;
- environmental conditions to be used, e.g., streamflow conditions, precipitation and temperature inputs, tidal fluctions;
- the specific objectives for the analysis and the priority of questions to be answered;
- the number of scenarios to be analyzed, i.e., combinations of values of exogenous variables; and
- the criteria for evaluating strategies and the relative weights of those criteria.

Whatever the context, SUTA is a necessary and critical part of analysis for marine resources management. It is the process which determines what types of analyses to undertake, in what order, to produce the desired information in relation to the questions to be answered, in the time available, for any given context in which the analysis is being made. The output of SUTA includes the detailing of the specific characteristics of the analysis and of the work schedule for completing the analysis in the time prescribed, i.e., when the information is needed for the decision process.

Segment 2: Estimating demands for products and services

The demands for products and services which can be produced from a given marine resources region are related to the levels of economic activities onshore and to international trade. Estimating the demands for products and services from a given region begins with demographic and economic projections over time at the national level, final demand associated therewith by various sectors, and "eventually" the qualities of various raw materials, e.g., manganese ore, sand and gravel, crude petroleum, needed to meet the input requirements of the activities producing the products and services. Demands at the national level plus relevant international demands must then be translated into projections of demands on the given marine resources region.

Segment 3: Analyzing activities and extra regional sources of wastes

Segment 3 is comprised of two parts. One involves the analysis of the activities producing the goods and services in the region in terms of the factors which influence both the choice of technology and the quantities of wastes generated and discharged. The other involves the analysis of extra regional "combined" sources of inputs of wastes to the region, namely, atmosphere, ocean currents, and rivers. To the extent possible, the types of activities and/or the specific geographic region associated with these combined sources should be identified.

Segment 4: Analyzing natural systems

Analyzing natural systems involves the estimation of the time and spatial pattern of ambient water quality and of effects on ecosystem components, e.g., various trophic levels, of the time and spatial pattern of discharges of wastes estimated in segment 3. The natural systems models used may be simple or complex.

Segment 5: Formulating and analyzing strategies

In chapter 7 the components of management strategies were defined. Typically many strategies can be formulated to reach any given objective formulated in the SUTA segment. To select among the possible strategies, consequences should be estimated in relation to the multiple criteria selected by the decision makers, examples of which were shown in table 7-5.

Segment 6: Evaluating strategies

Decision-makers use multiple criteria in making decisions. These criteria are specified in SUTA and the consequences of management strategies in terms of these criteria are estimated in segment 5. In addition, relative weights are attached to the criteria in the decision-making process. Assigning such relative weights to the individual criteria comprises an activity which is the responsibility of the decision makers, not of the analysts. The analyst, however, plays an important role in providing the motivation and the framework for making explicit and using these weights.

The explicit application of relative weights yields a rating of proposed strategies. This process is called evaluation. Analytical methods exist to contribute to the evaluation of strategies, such as cost benefit analysis and multi-criteria evaluation methods.

Segment 7: Presenting results

The types and amounts of data and the formats used to present the data will vary in relation to the questions being asked, the level of detail of the analysis, the structure of decision making in the region, and the characteristics of the region.

Not only the results of the analysis must be presented, but also information on: (1) selection and definition of the characteristics of the system considered; (2) selection and definition of goals and objectives; (3) selection and definition of mixes of activities; (4) selection and description of analytical methods and computational procedures, paying special attention to the underlying assumptions and the accuracy of the results; (5) selection and description of criteria used for evaluation; and (6) discussion of the data presented. Such information is helpful to the decision makers in making their decisions. In general it can be said that it should be considered a major task of the framework for analysis outlined in this document, that the selection procedure which leads to the finally presented promising strategies should be made explicit in a clear and understandable way.

Appendix B: EC cadmium directive

THE COUNCIL OF THE EUROPEAN COMMUNITIES

Having regard to the Treaty establishing the European Economic Community, and in particular Articles 100 and 235 thereof,

Having regard to Directive 76/464/EEC of 4 May 1976 on pollution caused by certain dangerous substances discharged into the aquatic environment of the Community[1], and in particular Articles 6 and 12 thereof,

Having regard to the proposal from the Commission[2],

Having regard to the opinion of the European Parliament[3],

Having regard to the opinion of the Economic and Social Committee[4],

Whereas, in order to protect the aquatic environment of the Community against pollution by certain dangerous substances, Article 3 of Directive 76/464/EEC introduces a system of prior authorization laying down emission standards for discharges of the substances in List I in the Annex thereto; whereas Article 6 of the said Directive provides that limit values shall be laid down for such emission standards and also quality objectives for the aquatic environment affected by discharges of these substances;

Whereas cadmium and its compounds are included in List I;

Whereas the Member States are required to apply the limit values except in the cases where they may employ quality objectives;

Whereas, since pollution due to the discharge of cadmium into water is caused by a large number of industries, it is necessary to lay down specific limit values according to the type of industry concerned and

---

1) OJ No. L 129, 18-5-1976, p. 23
2) OJ No. C 118, 21-5-1981, p. 3
3) OJ No. C 334, 20-12-1982, p. 138
4) OJ No. C 230, 10-9-1981, p. 22

to lay down quality objectives for the aquatic environment into which cadmium is discharged by such industries;

Whereas at the present time it is not, however, possible to establish limit values for discharges arising from the manufacture of phosphoric acid and phosphatic fertilizer from phosphatic rock;

Whereas the purpose of the quality objectives must be to eliminate cadmium pollution of the various parts of the aquatic environment which might be affected by cadmium discharges;

Whereas such quality objectives must be laid down expressly for this purpose and not with the intention of establishing rules pertaining to consumer protection or to the marketing of products from the aquatic environment;

Whereas a specific monitoring procedure should be laid down to enable Member States to demonstrate that the quality objectives are being complied with;

Whereas provision should be made for the monitoring by Member States of the aquatic environment affected by the aforesaid cadmium discharges with a view to effective implementation of this Directive; whereas Article 6 of Directive 76/464/EEC does not provide for the powers to introduce such monitoring; whereas, since the specific powers have not been provided for in the Treaty, Article 235 thereof should be invoked;

Whereas it is important that the Commission forward to the Council, every five years, a comparative assessment of the implementation of this Directive by Member States;

Whereas, since groundwater is the subject of Directive 80/68/EEC[1)], it is excluded from the scope of this Directive;

Whereas the level of industrialization is very low in Greenland because of the overall situation of the island, and in particular the fact that it is sparsely populated, its considerable size and its special geographical position; whereas, therefore, this Directive should not apply to Greenland,

HAD ADOPTED THIS DIRECTIVE:

Article 1

1. This Directive:
   - in pursuance of Article 6 (1) of Directive 76/464/EEC, lays down limit values for emission standards for cadmium in discharges from industrial plants as defined in Article 2 (e) hereof,
   - in pursuance of Article 6 (2) of Directive 76/464/EEC, lays down quality objectives for cadmium in the aquatic environment,

---

1) OJ No. L 20, 26-1-1980, p. 43

- in pursuance of Article 6 (4) of Directive 76/464/EEC, lays down the time limits for compliance with the conditions specified in the authorizations granted by the competent authorities of Member States in respect of existing discharges,
- in pursuance of Article 12 (1) of Directive 76/464/EEC lays down the reference methods of measurement enabling the cadmium content in discharges and in the aquatic environment to be determined,
- in pursuance of Article 6 (3) of Directive 76/464/EEC, establishes a monitoring procedure,
- requires Member States to cooperate with one another in the case of discharges affecting the waters of more than one Member State.

2. This Directive applies to the waters referred to in Article 1 of Directive 76/464/EEC with the exception of groundwater.

Article 2

a) 'cadmium' means:
   - the chemical element cadmium.
   - the cadmium contained in any of its compounds;

b) 'limit values' means the values specified in Annex I;
c) 'quality objectives' means the requirements specified in Annex II;
d) 'handling of cadmium' means any industrial process involving the use or production of cadmium, or any other process in which the presence of cadmium is inherent;
e) 'industrial plant' means any plant at which cadmium or any substance containing cadmium is handled;
f) 'existing plant' means an industrial plant which is operational on the date of notification of this Directive;
g) 'new plant' means:
   - an industrial plant which has become operational after the date of notification of this Directive,
   - an existing industrial plant whose cadmium-processing capacity has been substantially increased after the date of notification of this Directive.

Article 3

1. The limit values, the time limits by which they must be complied with and the monitoring procedure for discharges are laid down in Annex I.

2. The limit values shall normally apply at the point where waste waters containing cadmium leave the industrial plant.

   When waste waters containing cadmium are treated outside the industrial plant at a treatment plant intended for the removal of cadmium, the Member State may permit the limit values to be applied at the point where the waste waters leave the treatment plant.

3. The authorizations referred to in Article 3 of Directive 76/464/EEC must contain provisions at least as stringent as those in Annex I

to this Directive, except where a Member State is complying with Article 6 (3) of Directive 76/464/EEC on the basis of Annexes II and IV to this Directive.

Authorizations shall be reviewed at least every four years.

4. Without prejudice to their obligations arising from paragraphs 1, 2 and 3 and to the provisions of Directive 76/464/EEC, Member States may grant authorizations for new plants only if those plants apply the standards corresponding to the best technical means available when that is necessary for the elimination of pollution in accordance with Article 2 of the said Directive or for the prevention of distortion of competition.

   Whatever method it adopts, the Member State shall, where for technical reasons the intended measures do not correspond to the best technical means available, provide the Commission, before any authorization, with evidence in support of these reasons.

   The Commission shall forward this evidence to the other Member States immediately and shall send all Member States a report as soon as possible giving its opinion on the derogation referred to in the second subparagraph. If necessary, it shall at the same time submit appropriate proposals to the Council.

5. The reference method of analysis to be used in determining the presence of cadmium is given in Annex III (1). Other methods may be used provided that the limits of detection, precision and accuracy of such methods are at least as good as those laid down in Annex III (1). The accuracy required in the measurement of effluent flow is given in Annex III (2).

Article 4

The Member States concerned shall be responsible for monitoring the aquatic environment affected by industrial discharges.

In the case of discharges affecting the waters of more than one Member State, the Member States concerned shall cooperate with a view to harmonizing monitoring procedures.

Article 5

1. The Commission shall make a comparative assessment of the implementation of this Directive by Member States on the basis of information supplied to it by them pursuant to Article 13 of Directive 76/464/EEC at its request, which it must submit case by case. The information concerned shall, in particular, comprise:

   - details of authorizations laying down emission standards for discharges of cadmium,
   - the results of the inventory of cadmium discharged into the waters referred to in Article 1 (2),

- the results of measurements made by the national network set up to determine concentrations of cadmium.

2. The Commission shall forward the comparative assessment referred to in paragraph 1 to the Council every five years, and for the first time four years after notification of this Directive.

3. In the event of a change in scientific knowledge relating principally to the toxicity, persistence and accumulation of cadmium in living organisms and sediments, or in the event of an improvement in the best technical means available, the Commission shall submit appropriate proposals to the Council with the aim of reinforcing, if necessary, the limit values and the quality objectives or of establishing new limit values and new quality objectives.

Article 6

1. Member States shall bring into force the measures necessary to comply with this Directive within two years following its notification. They shall forthwith inform the Commission thereof.

2. Member States shall communicate to the Commission the text of the provisions of national law which they adopt in the filed governed by this Directive.

Article 7

This Directive shall not apply to Greenland.

Article 8

This Directive is addressed to the Member States.

Done at Brussels, 26 September 1983.

For the Council
The President
C. Simitis

ANNEX I

Limit values, time limits fixed for compliance with these values and monitoring procedures to be applied to discharges

1. Limit values and time limits

Industrial sector[1]	Unit of measurement	Limit values which must be complied with as from	
		1-1-1986	1-1-1989[2]
1. Zinc mining, lead and zinc refining, cadmium metal and non-ferrous metal industry	Milligrams of cadmium per litre of discharge	0,3[3]	0,2[3]
2. Manufacture of cadmium compounds	Milligrams of cadmium per litre of discharge	0,5[3]	0,2[3]
	Grams of cadmium discharged per kilogram of cadmium handled	0,5[4]	[5]
3. Manufacture of pigments	Milligrams of cadmium per litre of discharge	0,5[3]	0,2[3]
	Grams of cadmium discharged per kilogram of cadmium handled	0,3[4]	[5]
4. Manufacture of stabilizers	Milligrams of cadmium per litre of discharge	0,5[3]	0,2[3]
	Grams of cadmium discharged per kilogram of cadmium handled	0,5[4]	[5]
5. Manufacture of primary and secondary batteries	Milligrams of cadmium per litre of discharge	0,5[3]	0,2[3]
	Grams of cadmium discharged per kilogram of cadmium handled	1,5[4]	[5]
6. Electroplating[6]	Milligrams of cadmium per litre of discharge	0,5[3]	0,2[3]
	Grams of cadmium discharged per kilogram of cadmium handled	0,3[4]	[5]
7. Manufacture of phosphoric acid and/or phosphatic fertilizer from phosphatic rock[7]		--	--

1) Limit values for industrial sectors not mentioned in this table will, if necessary, be fixed by the Council at a later stage. In the meantime the Member States will fix emission standards for cadmium discharges autonomously in accordance with Directive 76/464/EEC. Such standards must take into account the best technical means available and must not be less stringent than the most nearly comparable limit value in this Annex.
2) On the basis of experience gained in implementing this Directive, the Commission will, pursuant to Article 5 (3), submit in due course to the Council proposals for fixing more restrictive limit values with a view to their coming into force by 1992.
3) Monthly flow-weighted average concentration of total cadmium.
4) Monthly average.
5) It is impossible for the moment to fix limit values expressed as load. If need be, these values will be fixed by the Council in accordance with Article 5 (3) of this Directive. If the Council does not fix any limit values, the values expressed as load given in column '1-1-1986' will be kept.
6) Member States may suspend application of the limit values until 1 January 1989 in the case of plants which discharge less than 10 kg of cadmium a year and in which the total volume of the electroplating tanks is less than 1,5 m^3, if technical or administrative considerations make such a step absolutely necessary.
7) At present there are no economically feasible technical methods for systematically extracting cadmium from discharges arising from the production of phosphoric acid and/or phosphatic fertilizers from phosphatic rock. No limit values have therefore been fixed for such discharges. The absence of such limit values does not release the Member States from their obligation under Directive 76/464/EEC to fix emission standards for these discharges.

2. Limit values expressed as concentrations which in principle must not be exceeded are given in the above table for the industrial sectors in sections 2, 3, 4, 5 and 6. In no instance may limit values expressed as maximum concentrations be greater than those expressed as maximum quantities divided by water requirements per kilogram of cadmium handled. However, because the concentration of cadmium in effluents depends on the volume of water involved, which differs for different processes and plants, the limit values, expressed in terms of the quantity of cadmium discharged in relation to the quantity of cadmium handled, given in the above table must be complied with in all cases.

3. The daily average limit values are twice the corresponding monthly average limit values given in the above table.

4. A monitoring procedure must be instituted to check whether the discharges comply with the emission standards which have been fixed in accordance with the limit values laid down in this Annex.

   This procedure must provide for the taking and analysis of samples and for measurement of the flow of the discharge and the quantity of cadmium handled.

   Should the quantity of cadmium handled be impossible to determine, the monitoring procedure may be based on the quantity of cadmium that may be used in the light of the production capacity on which the authorization was based.

5. A sample representative of the discharge over a period of 24 hours will be taken. The quantity of cadmium discharged over a month must be calculated on the basis of the daily quantities of cadmium discharged.

   However, a simplified monitoring procedure may be instituted in the case of industrial plants which do not discharge more than 10 kg of cadmium per annum. In the case of industrial electroplating plants, a simplified monitoring procedure may only be instituted if the total volume of the electroplating tanks is less than 1,5 m^3.

ANNEX II

Quality objectives

For those Member States which apply the exception referred to in Article 6 (3) of Directive 76/464/EEC, the emission standards which Member States must establish and ensure are applied, pursuant to Article 5 of that Directive, will be fixed so that the appropriate quality objective or objectives from among those listed below is or are complied with in the area affected by discharges of cadmium. The competent authority shall determine the area affected in each case and shall select from among the quality objectives listed in paragraph 1 the objective or objectives that it deems appropriate having regard to the intended use of the area affected, while taking account of the fact that the purpose of this Directive is to eliminate all pollution.

1. The following quality objectives[1)], which will be measured sufficiently close to the point of discharge, are fixed, with the object of eliminating pollution within the meaning of Directive 76/464/EEC and pursuant to Article 2 of that Directive[2)]:

1.1. The total cadmium concentration in inland surface waters affected by discharges must not exceed 5 µg/litre.

1.2. The concentration of dissolved cadmium in estuary waters affected by discharges must not exceed 5 µg/litre.

1.3. The concentration of dissolved cadmium in territorial waters and in internal coastal waters other than estuary waters affected by discharges must not exceed 2,5 µg/litre.

1.4. In the case of waters used for the abstraction of drinking water, the cadmium content must conform to the requirements of Directive 75/440/EEC[3)].

2. In addition to the above requirements, cadmium concentration must be determined by the national network referred to in Article 5 and the results compared with the following concentrations[2)]:

2.1. In the case of inland surface waters, a total cadmium concentration of 1 µg/litre.

2.2. In the case of estuary waters, a dissolved cadmium concentration of 1 µg/litre.

---

1) The cadmium concentrations indicated in 1.1, 1.2 and 1.3 are the minimum requirements necessary to protect aquatic life.

2) With the exception of quality objective 1.4, all concentrations relate to the arithmetic mean of the results obtained over one year.

3) Directive 75/440/EEC concerns the quality required of surface water intended for the abstraction of drinking water in the Member States (OJ No. L 194, 25-7-1975, p. 26). It provides for a mandatory cadmium value of 5 µg/litre on the basis of 95% of the samples taken.

2.3. In the case of territorial and internal coastal waters, other than estuary waters, a dissolved cadmium concentration of 0,5 μg/litre.
If these concentrations are not complied with at any one of the points on the national network, the reasons must be reported to the Commission.

3. The concentration of cadmium in sediments and/or shellfish, if possible of the species Mytilus edulis, must not increase significantly with time.

4. Where several quality objective are supplied to waters in an area, the quality of the waters must be sufficient to comply with each of those objectives.

ANNEX III

Reference methods of measurement

1. The reference method of analysis used for determining the cadmium content of waters, sediments and shellfish is atomic absorption spectrophotometry after preservation and suitable treatment of the sample.

   The limits of detection[1)] must be such that the cadmium concentration can be measured to an accuracy of ± 30% at the following concentrations:

   - in the case of discharges, one-tenth of the maximum permitted concentration of cadmium specified in the authorization,
   - in the case of surface water, 0,1 µg/litre or one-tenth of the cadmium concentration specified in the quality objective, whichever is the greater,
   - in the case of shellfish, 0,1 mg/kg, wet weight,
   - in the case of sediments, one-tenth of the cadmium concentration in the sample or 0,1 mg/kg, dry weight, with drying being carried out between 105 and 110°C at constant weight, whichever value is the greater.

2. Flow measurement must be carried out to an accuracy of ± 20%.

---

1) The definitions of these terms are given in Council Directive 79/869/EEC of 9 October 1979 concerning the methods of measurement and frequencies of sampling and analysis of surface water intended for the abstraction of drinking water in the Member States (OJ No. L 271, 29-10-1979, p. 44)

ANNEX IV

Monitoring procedure for quality objectives

1. For each authorization granted in pursuance of this Directive, the competent authority will specify the restrictions, monitoring procedure and time limits for ensuring compliance with the quality objective(s) concerned.

2. In accordance with Article 6 (3) of Directive 76/464/EEC, the Member State will, for each quality objective chosen and applied, report to the Commission, on:

   - the points of discharge and the means of dispersal,
   - the area in which the quality objective is applied,
   - the location of sampling points,
   - the frequency of sampling,
   - the methods of sampling and measurement,
   - the results obtained.

3. Samples must be sufficiently representative of the quality of the aquatic environment in the area affected by the discharges, and the frequency of sampling must be sufficient to show any changed in the aquatic environment, taking into account, in particular, natural variations in the hydrological regime.

IFIAS

The International Federation of Institutes for Advanced Study is an association of 35 leading research institutes which collaborate to address major global problems of long-term importance in environment, economy and science and technology. IFIAS research programmes are interdisciplinary, seeking to advance understanding of complex systems for the improved management in a rapidly changing world with an uncertain future. IFIAS stands for the more effective and consistent use of scientific understanding in world councils, and for the adoption of long-term strategic thinking.

Chairman:
Sir Hermann Bondi
Master of Churchill College
Masters Lodge
Churchill College
Cambridge CB3 ODS
England

IFIAS
Director: Ian Burton
Haultain Building
170 College St.
TORONTO, Ontario
Canada

IFIAS-Maastricht
Project Development Office
Director: Frank Rijsberman
Onze Lieve Vrouweplein 21
6211 HE MAASTRICHT
The Netherlands

MEMBER INSTITUTES

Athens Center of Ekistics, Athens, Greece
Centro de Investigacion y de Estudios Avanzados del IPN, CINVESTAV, Mexico D.F., Mexico
El Colegio de Mexico A.C., Mexico D.F., Mexico
Delft Hydraulics, Delft, The Netherlands
Food Research Institute, Stanford, USA
Global Studies Center, Arlington, USA
The Graduate Institute of International Studies, Geneva, Switzerland
Institut National de la Recherche Scientifique, Sainte-Foy, Canada
Institute for Environmental Studies, Toronto, Canada
Institute for Futures Studies, Copenhagen, Denmark
Institute for Studies on Research and Scientific Documentation, Rome, Italy
Institute for World Economics of the Hungarian Academy of Sciences, Budapest, Hungary
Instituto de Biofisica, Rio de Janeiro, Brazil
Instituto Brasileiro de Economia, Rio de Janeiro, Brazil
Instituto de Ciencias del Hombre, I.C.H., Madrid, Spain
Instituts Internationaux de Physique et de Chimie, Brussels, Belgium
International Centre of Insect Physiology and Ecology, ICIPE, Nairobi, Kenya
International Centre for Theoretical Physics, ICTP, Trieste, Italy
The Jacob Blaustein Institute for Desert Research, Israel
The Japan Economic Research Center, JERC, Tokyo, Japan
Marga Institute, Colombo, Sri Lanka
National Institute for Research Advancement, NIRA, Tokyo, Japan
National Research Center for Science and Technology for Development, NRCSTD, Beijing, People's Republic of China
National Research Centre, Dokki-Cairo, Egypt
Niels Bohr Institute, Copenhagen, Denmark
Research Institute, Dhahran, Saudi Arabia
Research Policy Institute, Lund, Sweden
Royal Scientific Society, Amman, Jordan
Science Policy Research Unit, Brighton, England
Tata Institute of Fundamental Research, Bombay, India
University City Science Center, Philadelphia, USA
University Corporation for Atmospheric Research, UCAR, Boulder, USA
The Weizmann Institute of Science, Rehovot, Israel
Winrock International Institute for Agricultural Development, Morrilton, USA
Woods Hole Oceanographic Institution, WHOI, Woods Hole, USA

CORPORATE AFFILIATES

AKZO NV, Arnhem, The Netherlands
Atlas Copco AB, Stockholm, Sweden
DSM, Heerlen, The Netherlands
ENEA, Rome, Italy
Gist Brocades NV, Delft, The Netherlands
Oce-van der Grinten, Venlo, The Netherlands
Philips International, Eindhoven, The Netherlands
Royal Dutch Shell, Amsterdam, The Netherlands
Skandinaviska Enskilda Banken, Stockholm, Sweden
Svenska Handelsbanken, Stockholm, Sweden
Unilever, Vlaardingen, The Netherlands
VBB AB, Stockholm, Sweden
Volker-Stevin Wegen en Asfalt, Utrecht, The Netherlands

Secretariate Coastal Waters Management Programme, Delft Hydraulics, P.O. Box 177, 2600 MH Delft, The Netherlands

IFIAS RESEARCH SERIES:

Vol. 1. G.Hallsworth, *The Anatomy, Physiology and Psychology of Erosion,* John Wiley and Sons Ltd. Chichester, 1987.

Vol. 2. Rob Koudstaal, *Water Quality Management Plan, North Sea: Framework for Analysis,* A.A.Balkema, Rotterdam, 1987, Coastal Waters No.1.

Vol. 3. Lies Dekker, Blair T.Bower & Rob Koudstaal, *Management of Toxic Materials in an International Setting, A Case Study of Cadmium in the North Sea,* A.A.Balkema, Rotterdam, 1987, Coastal Waters No. 2.

IFIAS

international federation of institutes
for advanced study

delft hydraulics

ifias research series **coastal waters nr. 2**